D1572564

GEOTHERMAL ENERGY AS A SOURCE OF ELECTRIC POWER

Stanley L. Milora is a member of the Cold Vapor Technology Program, Oak Ridge National Laboratory, Oak Ridge, Tennessee.
Jefferson W. Tester is a member of the Geothermal Energy Group, Los Alamos Scientific Laboratory, Los Alamos, New Mexico.

GEOTHERMAL ENERGY AS A SOURCE OF ELECTRIC POWER
Thermodynamic and Economic Design Criteria

Stanley L. Milora
and
Jefferson W. Tester

The MIT Press
Cambridge, Massachusetts, and London, England

PUBLISHER'S NOTE
This format is intended to reduce the cost of publishing certain works in book form and to shorten the gap between editorial preparation and final publication. Detailed editing and composition have been avoided by photographing the text of this book directly from the authors' typescript.

Printed in the United States of America

Second printing, March 1977
Third printing, November 1977

Library of Congress Cataloging in Publication Data

Milora, Stanley L
Geothermal energy as a source of electric power.

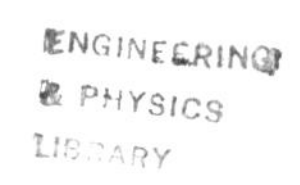

Bibliography: p.
Includes index.
1. Geothermal engineering. 2. Thermodynamics. 3. Geothermal resources.
I. Tester, Jefferson W., joint author. II. Title.
TK1055.M54 621.312'139 76-7008
ISBN 0-262-13123-4

CONTENTS

DEDICATED

TO

THERESA

AND

SUE

PREFACE

The current energy crisis has generated an increased interest in alternate sources of energy. Resources such as geothermal, solar, and ocean thermal are being actively considered for wide-spread development in the United States and many other countries throughout the world. All of these resources share the disadvantage of having an inherently low thermodynamic quality (temperature).

One technique of increasing their quality is to convert their heat into electric power. Unfortunately, the efficiency of this conversion process is limited by the temperature of the resource and by prevailing ambient conditions for heat rejection. Efficiencies for converting geothermal resources occurring at temperatures below 250°C to electricity are substantially less than those of fossil-fuel-fired or nuclear power plants. This creates strong economic incentives for optimizing the performance of any power conversion scheme involving these new resources.

New technologies are rapidly developing for improving the thermodynamic efficiency of power production toward its ideal limit as well as providing an economical process. For example, instead of using water as the working fluid, several hydrocarbons and their halogenated derivatives are being proposed. They have considerably different properties from water in the low-temperature region, perhaps making them more suitable working fluids.

As the energy program of this country evolved, we felt a need to investigate the technological aspects of producing electric power from low-temperature resources. This monograph focuses on the utilization of geothermal resources and attempts to provide a realistic framework where both thermodynamic conversion efficiency and economics are considered in detail. Our main purpose in preparing this manuscript is to discuss factors which we feel are important to the development of geothermal energy as a viable source of electric power. We expect it to complement the current work of the U.S. Geological Survey, the National Science Foundation, programs within the Energy Research and Development Administration, and the many industrial ventures into geothermal energy.

We are writing for a diverse audience since the geothermal field spans many disciplines. Hopefully, the material presented will be useful to engineers interested in power production as well as to geologists, geochemists, and geophysicists who may be more involved with resource development.

We would like to thank the people who contributed to this effort. The contributions made by the participants of the Massachusetts Institute of Technology School of Chemical Engineering Practice are appreciated. In particular, the efforts of R.M. Mayer, K.R. Landgraf, K.I. Kudrnac, R. Solares. A.S.Y. Ho, P.L. Jensen, and P.C. Ahrens were very helpful in the early period of this work. We also thank R.N. Lyon, M.C. Smith, D.W. Brown, S.E. Beall, L.C. Fuller, S. Combs, R.B. Duffield, K.E.

Nichols, R.M. Potter, R.D. McFarland, R.D. Foster, R.H. Hendron, C.H. Peterson, D.L. Hanson, S.M. Fleming, J.C. Rowley, and J.H. Altseimer for their interest and assistance. We are also grateful for the dedication of Barbara Ramsey, Marjorie Wilson, and Maxine Lewis in preparing the manuscript for publication. The financial support provided by the Geothermal Energy Research Division of the Energy Research and Development Administration is gratefully acknowledged. The work was performed under contract numbers W-7405-eng-26 and W-7405-eng-36.

1. SUMMARY AND SCOPE

Optimum utilization of a geothermal energy resource, in both an economic and thermodynamic sense, is strongly dependent on the characteristics of the geothermal fluid. In particular, temperature, pressure, composition, and liquid to vapor ratio are important in determining the best method and conditions for energy conversion. Because the temperatures of most geothermal fluids are low, producing electric power by a conventional Rankine or similar cycle is an inherently inefficient process, but one, nonetheless, which has received worldwide attention.[1-7]

Prior to considering various power conversion schemes, the geothermal resource itself is discussed in Chapter 2. The characteristics and magnitude of this country's natural hydrothermal resource as well as the artificially stimulated, dry hot rock resource are described briefly. The environmental aspects of geothermal energy exploitation are also explored.

As indicated by references 1-8, research and development activity in the area of geothermal power cycle development has increased markedly in the past few years. Water is frequently used as a working fluid in the conversion cycle, particularly where natural steam is available, but it might not be the most economic or efficient choice. Binary-fluid cycles employing working fluids other than water are being developed as alternatives to single- and multiple-flashing systems currently in use in various parts of the world (Cerro Prieto, Mexico and Wairakei, New Zealand)[8].In particular, work by Jonsson, Taylor, and Charmichael[6], Anderson[1], and Cortez, Holt, and Hutchinson [4] represent pioneering efforts in binary-fluid cycle development. We have concentrated our efforts on related alternative schemes involving non-aqueous working fluids.

In order that the economic factors pertinent to producing electric power from liquid-dominated or dry hot rock geothermal fluids ranging from 100 to 300°C can be discussed, a thermodynamic framework is developed in which resource utilization criteria are emphasized for binary-fluid and flashing systems. In Chapter 3, cycle and resource utilization efficiencies are introduced to illustrate the limitations of producing useful work (electricity) from a geothermal fluid which varies in temperature as heat is removed. And because the thermodynamic properties of the working fluid are important, semi-empirical equations are presented for evaluating properties required in the cycle calculations. A modified form of the Martin-Hou equation of state with 21 parameters is used with empirical equations for the ideal-gas-state, constant pressure heat capacity, vapor pressure, and liquid density to calculate desired properties such as enthalpy and entropy.

In Chapter 4, detailed cycle calculations were performed with seven representative working fluids, selected to provide a range of molecular properties which are important in determining cycle performance with a given geothermal resource temperature. Various cycle configurations of 100 MW(e) capacity utilizing supercritical or subcritical Rankine and Brayton gas cycles are discussed and compared to single and multiple-stage flashing systems. The use of two secondary fluids with one geothermal resource in a topping/bottoming or dual cycle arrangement is also considered. Figure 1 shows schematics for these systems. Several newer techniques still under development which involve direct two-phase conversion, e.g., the helical screw expander or the total flow turbine [9] are not compared to the other systems.

Thermodynamic optimum conditions with respect to resource utilization were calculated by varying the system operating pressure of the binary-fluid cycle at a fixed geothermal resource temperature. To facilitate calculations computer computations were used in our cycle analyses. Flashing systems were also optimized for similar resource conditions. In each case a utilization efficiency, η_u, was determined which related the actual electrical work produced by the cycle to the maximum work (or availability) possible with specified geothermal source and heat rejection temperatures. Turbine and pump efficiencies and minimum acceptable temperature differences (pinch points) were specified for each working fluid considered. An irreversibility analysis of cycle performance was also developed to illustrate how individual cycle components affect η_u. By examining the behavior of the seven representative fluids which included ammonia (NH_3), isobutane (i-C_4H_{10}), R-32 (CH_2F_2), RC-318 (C_4F_8), R-114 ($C_2Cl_2F_4$), R-115 (C_2ClF_5), and R-22 ($CHClF_2$), several generalizations were proposed that could be useful in screening potential fluids. By plotting optimum η_u's for each fluid as a function of resource temperature (100 to 300°C), one observes a characteristic maximum η_u at a particular resource temperature. These maxima of η_u are different for each fluid but generally range from 60 to 70%. We were able to relate the temperature, T*, corresponding to maximum η_u to the ideal gas state constant pressure heat capacity C_p^* by expressing T* as an amount of superheat above the critical point, T_c

$$T^* - T_c = \frac{790}{C_p^*/R}$$

In a qualitative sense, this degree of superheat could be correlated to the molecular properties of the working fluid.

Topping/bottoming or dual cycle arrangements utilizing a dry hot rock resource at 280°C were optimized using techniques similar to those used for single fluid cycles. Their performance ($\eta_u \cong 61\%$) was competitive but not superior to that of a single fluid cycle. Brayton gas cycles, on the other hand, have inherently low efficiencies because of the temperatures involved with geothermal systems.

Because turbine size and thus cost vary considerably when low pressure steam or higher vapor pressure organic compounds are used as working fluids, criteria applicable to turbines and pumps are discussed in detail in Chapter 5. Basically, a similarity analysis of turbine performance was developed to assist in selecting operating conditions and sizes which would result in optimum efficiencies. Equations and graphs are presented which interrelate turbine rotational speed, stage enthalpy drop, blade pitch diameter, and stage volumetric flow capacity. In addition, a figure of merit was devised using a generalized approach to turbine exhaust flow capacity at subsonic flow conditions. Exhaust flow capacity was related to the fluid's molecular weight, critical pressure, saturated vapor specific volume at the heat rejection temperature, and latent heat of vaporization. Figures of merit are presented for over twenty compounds ranging in properties from ammonia and isobutane to water. The smaller turbine sizes resulting from fluids having a low molecular weight, high critical pressure, and high vapor-phase energy density (ratio of latent heat to vapor specific volume) are evident. For example, with a turbine exhaust temperature of 26.7°C(80°F) an ammonia turbine would be approximately 150 times smaller than a steam turbine of the same power output. This is caused primarily by the differences in vapor specific volume at 26.7°C. A generalized correlation for feed pump power requirements is also developed.

The economics of power cycles are covered in Chapter 6. An economic model is developed which bases the total capital investment on the costs of the geothermal production and re-injection wells and major pieces of equipment, including pumps, turbines, generators, heat exchangers, and condensers. Both natural hydrothermal and dry hot rock geothermal resources are considered. Direct and indirect cost factors are introduced to obtain estimates of installed costs from purchased costs. Separate cost equations or correlations are presented for each of the major components. For example, turbine sizes determined by the equations of Chapter 5 can be used in a parametric equation to estimate turbogenerator costs. A discussion of the factors influencing heat exchanger and condenser sizes is also included with the cost data. In particular, correlations for estimating heat transfer coefficients including fouling effects and the effect of operating pressure are included.

Because of the numerous variables that need to be specified before costs can be estimated, several specific cases were examined. A 100 MW(e) capacity was selected. A 150°C liquid- dominated resource with a two-stage flashing and an R-32 (CH_2F_2) binary-fluid cycle and a 250°C dry hot rock resource with ammonia as the working fluid were considered initially. In each case, well flow rate and depth were specified and reinjection well costs included. Economic optima were determined by varying the operating pressure of the binary-fluid cycle for fixed heat exchange conditions. Flashing system optima were calculated by varying the flashing stage pressure or temperature.

Cost estimates of this type are not only very site specific in the sense that they depend on well flow rate, geothermal fluid temperature, and geothermal temperature gradient, but they also depend heavily on the assumptions made concerning the process equipment and well drilling costs. We have tried to be conservative in our estimates, particularly with respect to well costs, and consequently total capital investments are high. For the 150°C liquid-dominated resource, costs range between $1400/kW to $2200/kW with between 45 to 80% of that figure invested in the wells depending on cycle choice. This case was for a geothermal gradient of 50°C/km which is far from the anomalously high gradients of 200°C/km found in certain regions of the world. Using a 50°C/km gradient but drilling to obtain 250°C fluids in a dry hot rock system with a well flow rate 3 times higher, costs drop to $600/kW.

At this point, the controlling effects that well flow rates, fluid temperatures, and geothermal gradients have on the economics should be evident. The effect of resource temperature on generating costs was determined by optimizing an R-32 cycle at several geothermal fluid temperatures between 130 and 250°C assuming a constant geothermal gradient of 50°C/km and a fixed well flow rate. There was a distinct minimum in the cost curve at a particular temperature. In other words, for a given set of resource and power plant conditions there is an optimum depth for drilling.

These results were expanded into a generalized cost model for preliminary estimating purposes. In this model, installed generating cost is expressed parametrically as a function of well flow rate (45 to 225 kg/sec), geothermal fluid temperature (100 to 300°C), and geothermal gradient (20 to 200°C/km) using a binary-fluid cycle for power production.

Given our conservative economic approach, geothermal generating costs still appear competitive with the present escalated fossil-fuel and nuclear generating costs, and should be given serious consideration as one alternate energy source.

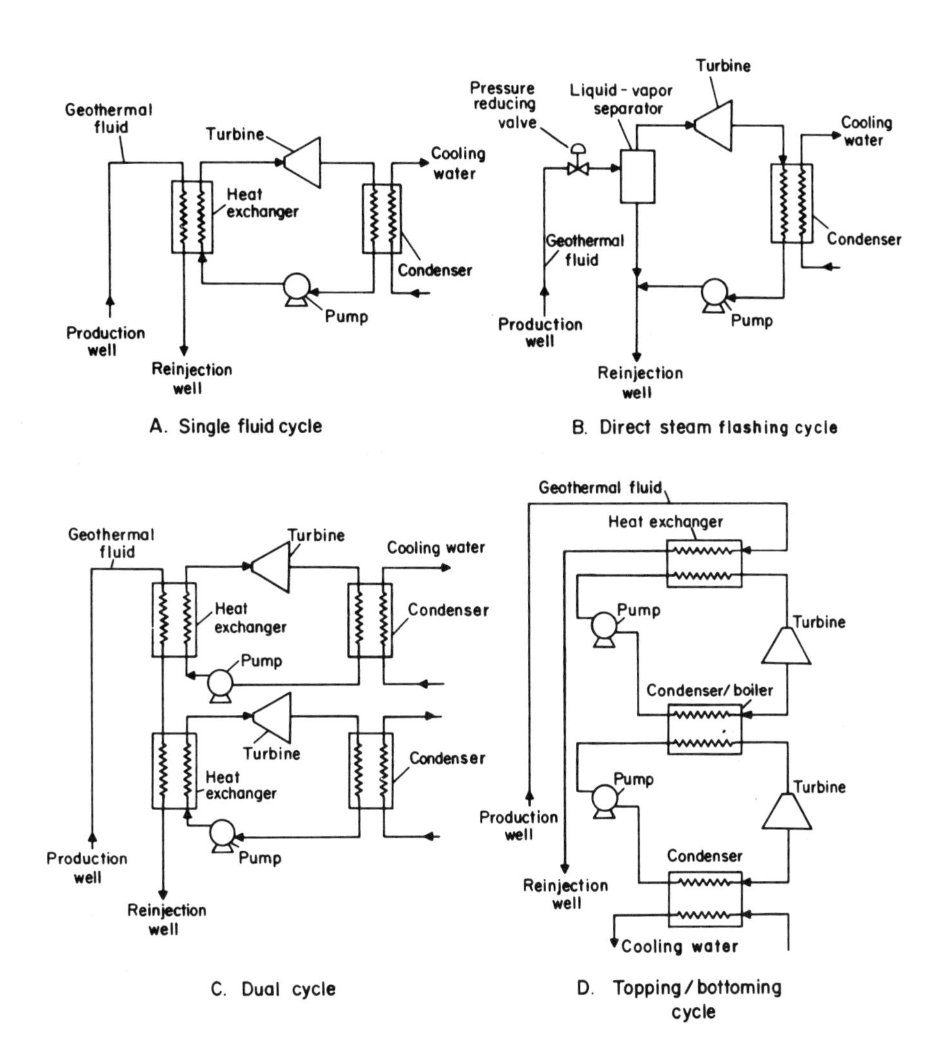

Fig. 1
Schematic of cycle configurations for geothermal power production.

2. GEOTHERMAL RESOURCES

2.1 Resource Characteristics

With a volume of 10^{12} km^3 of mostly molten rock, the earth represents a potentially inexhaustable source of energy. Typical surface manifestations of geothermal energy are evident in the form of hot springs, geysers, and fumaroles, not to overlook active volcanoes.[8] Unfortunately its inaccessibility, except in a relatively few areas of the world, has severely limited the exploitation of geothermal energy.

Currently, conventional drilling techniques are used to reach natural underground reservoirs (aquifers) that contain hot water and/or steam. These geothermal fluids are carried to the surface where they are either used directly for space or process heating purposes or to produce electricity from a vapor turbine cycle. Natural geothermal systems can be divided into four categories: (1) vapor-dominated (dry steam), (2) liquid dominated (superheated water), (3) geopressured reservoirs, and (4) lavas and magmas.[8]

Although dry steam fields are relatively rare, the Italian fields at Larderello and Mt. Amiata and the U.S. field at The Geysers in California, are producing over 700 MW(e) of electricity. Vapor-dominated fluids are advantageous for power production because they are usually available at relatively high temperature and high pressure (180°C, 7.8 bars (114 psia) at The Geysers [5,8]) and the superheated steam can be used directly to drive turbines.

The more commonly occurring liquid-dominated systems present a complex utilization problem in that a reasonably high-pressure vapor phase must be created for power conversion in a conventional turbogenerator unit. Fluid can be partially vaporized by flashing it to a lower pressure, with the vapor then injected into a suitable turbine to produce power.[5] Or it can exchange heat to another (secondary) fluid vaporizing at a lower temperature (e.g., isobutane or a halogenated hydrocarbon [Freon]) which in turn, is injected into a turbine, condensed, and pumped in a continuous closed cycle.[1-4] A disadvantage of direct steam flashing is that multiple flashing steps are required to attain high conversion efficiencies. Only a small fraction of steam is produced with a single flashing stage (about 10% for a 150°C source). Successive flashing improves efficiency but requires complex turbine design. Furthermore, the relatively high specific volume of steam at these lower temperatures results in large, expensive turbine exhaust areas. Both systems have to reinject or discard all or part of the geothermal fluid. Steam flashing, however, does require less surface equipment (such as heat exchangers and pumps) for power conversion than do secondary or alternate fluid cycles. Except for a few small systems (Kamchatka, USSR [7]), flashing plants are the only systems presently developed for liquid-dominated resources.

Liquid-dominated systems vary widely in terms of the available temperatures and pressures of geothermal fluids. For power production, temperatures of at least 100°C are desirable when coupled to sink (heat rejection) temperatures of approximately 25°C. Fluid temperatures as high as 300°C have been observed in the Imperial Valley of California and presently comprise an upper limit.

Geopressured reservoirs such as those of the Gulf Coast, from Mexico to Mississippi, contain moderately hot water (150 to 180°C) under extremely high pressures (270 to 400 bars).[8] However, utilization of this resource has been limited by engineering problems associated with drilling into such formations and extracting useful amounts of energy. Lavas and molten magmas are another potentially useful energy source; but controlled energy extraction is only in the formative research stages at this point.

Artifically stimulated geothermal resources are also under development in western sections of the U.S. One concept consists of drilling into dry hot rock and creating a geothermal reservoir by hydraulically fracturing the rock.[8,10] Water is then circulated through the fractured zone to remove heat and pumped to the surface. Additional surface area for *in situ* downhole heat transfer may also be created by thermal stress cracking which will greatly enhance the lifetime of the reservoir.[8] Energy removal on the surface may utilize direct steam flashing or a binary-fluid cycle for power production.

Another important factor in the design of geothermal power plants for both natural and artificially stimulated resources is the chemical composition of the fluid. Frequently, large quantities of dissolved minerals, partcularly silica (SiO_2) and calcium carbonate ($CaCO_3$) are present and they create scaling, corrosion, and erosion problems for heat exchangers, turbines, and related surface equipment. These problems vary in magnitude from those created by geothermal water containing less than 100 ppm total dissolved solids to those from Imperial Valley water with up to 300,000 ppm.[8]

Estimating reservoir lifetime in terms of geothermal fluid production rates and fluid quality is very difficult but is extremely critical to optimal economic design of a power plant. Furthermore, the reinjection of spent fluid to alleviate surface disposal problems, to control subsidence, and help in sustaining reservoir lifetime is far from a proven technique. Communication between a production and reinjection well is not at all certain in the case of natural reservoirs. For this reason, maximum use should be made of such wells; the difference between the geothermal fluids' well head and reinjection temperature should be maximized to the limit governed by available ambient conditions. For artificially stimulated resources, communication will be established prior to production but reinjection should occur at the lowest practical temperatures to favor thermal stress cracking.

TABLE 1

GEOTHERMAL RESOURCES
(based on USGS Circular 726[11])

System Type	Identified Resource		Total Resource	Total Recoverable as Thermal Energy	Total Recoverable as Electricity	References
	Number	Q 10^{21}J(~10^{18}BTU)	Q 10^{21}J(~10^{18}BTU)	Q 10^{21}J(~10^{18}BTU)	MW(e)-Centuries	
Hydrothermal Convection						
Vapor-dominated	3	0.11	0.21			11(Table 26)
Liquid-dominated (T_{gf}>150°C)	63	1.55	~6.70	2.97[d]	46,000	11(Table 26)
Liquid-dominated (90°C<T_{gf}<150°C)	224	1.44	~5.86			11(Table 26)
Geopressured Reservoirs[a,b]	---	~45.7	~184	2.3[c]	38,140	11(Tables 28 and 29)
Hot Igneous						
Molten magmas	---	~54.4	~419	?	?	11(Table 26)
Crystallized, hot sections	---	~50.2				
Conduction-dominated (dry hot rock, 0-10 km, T>15°C)	---	~33488	~33488	335	8,500,000	Recoverable energy assumed at 1% - this work and 11(Table 26)
Totals		33641	34104	340	8,584,000	

[a]Includes heat content of the pore fluid, and the thermal equivalents of contained methane and mechanical pressure energy (Table 28, Ref. 11).

[b]Assuming plan 2 suggested by Papadopulos and Associates[11] is adopted, that is unrestricted subsidence, no mechanical energy recovered, only the Gulf Coast geopressured fluid used for recovery (25,850 wells).

[c]Only 1.5 Q converted to electricity at 8% efficiency.

[d]Only 1.52 Q converted to electricity at efficiencies ranging from 8 to 20%.

2.2 Magnitude of the Resource

The United States Geological Survey [11,12] (USGS) and others [13-16] have estimated the total geothermal resource of the U.S., but as White and Williams [11] point out, the uncertain characteristics of the resource itself make this task extremely problematic. Furthermore, the state of current technology and the prevailing economic conditions with respect to other more conventional energy sources are important factors which strongly influence the predictions of any assessment model. Our intent in this section is to summarize briefly the latest available information on geothermal resource assessment. Circular 726 of the USGS [11] addresses this topic directly and is our prime source of information.

The ground rules of the USGS assessment are established by White and Williams.[11] The geothermal resource base is defined to include all discovered and undiscovered energy as stored heat above 15°C to a 10 km depth in all 50 states. Geothermal resources are categorized as:[11]

(1) Submarginal — recoverable at costs greater than two times current costs.

(2) Paramarginal — recoverable at costs between one and two times current costs.

(3) Reserves — recoverable at costs competitive with current costs.

Hydrothermal convection systems (vapor- and liquid-dominated and geopressured reservoirs), hot igneous systems (lavas and magmas), and conduction-dominated areas (dry hot rock) are discussed in separate papers contained in Circular 726.

The estimated total and recoverable energy resource is presented in Table 1. These quantities can be placed in perspective when the total energy needs for the U.S. are considered. The quantity $Q = 10^{21}$ J = $\sim 10^{18}$ BTU = $\sim 1.6 \times 10^{11}$ barrels of oil = 3.7×10^{10} short tons of coal is frequently used in calculations of this type. The total energy consumed by the U.S. in 1972 was 0.07 Q and the current electrical production is about 380,000 MW(e). Table 1 shows that the estimated recoverable geothermal resource is substantial at 340 Q as thermal energy or at 8,584,000 MW(e)-centuries as electrical energy. Of course, there are uncertainties in any estimates of this type; but one point that still remains is that geothermal resources are large enough to have an impact on our energy economy. Furthermore, the estimates of the USGS are probably conservative because their calculations assume that only a small fraction of the total resource is actually recoverable.

2.3 Environmental Factors

In comparing the environmental impact of geothermal energy with that of other energy sources such as oil, gas, coal, and nuclear, as producers of electric power, more than just the impact of the power plant should be considered.[17] Bowen [18] points out the necessity of examining the total fuel cycle. Geothermal installations, like hydroelectric, solar, or wind systems, have the entire fuel cycle located at the power generating station. Nuclear and fossil fuel cycles are not localized in one area and have a large spectrum of environmental impacts.

Environmental effects during the preparation and operation phases of a geothermal system are likely to be very site specific and very disparate. The preparation phase consists of land acquisition, active drilling and reservoir development, surface plant construction, and plant startup. The operation phase consists of power generation and any additional drilling or reservoir development required to sustain performance. Environmental factors are treated separately in the sections that follow.

2.3.1 Land Use

Although geothermal power plants are relatively rare in comparison to more conventional fossil-fuel plants, evidence at both The Geysers area in California and the Larderello fields in Italy indicates that surface land utilized for wells and surface plumbing is compatible with other land uses such as farming and cattle grazing.[18] At the power plant site, much larger land areas will be required for a geothermal plant versus a fossil-fuel or nuclear plant of similar capacity. However, the entire fuel cycle of a fossil-fuel or nuclear system could encompass a substantially greater land use. Bowen [18] cites the mining, milling, refining, enrichment, conversion, and fabrication steps of the nuclear fuel cycle which would require movement of ~4,000,000 tons of uranium ore over the 30 yr lifetime of a 1000 MW(e) plant. A coal-fired plant of similar capacity would require movement of ~200,000,000 tons of material assuming a 2:1 ratio of overburden to coal. Although oil- or natural gas-fired plants might have a lower land use impact, reserves within the U.S. are diminishing and the number of new plants constructed of this type will be small.

2.3.2 Noise

Noise problems are centered around the drilling, well testing, and pressure reducing valves typically used in flashing systems such as those at Cerro Prieto, Mexico and Wairakei, New Zealand. Muffling devices [18] and silencers [19] are presently used to reduce noise levels.

2.3.3 Subsidence and Seismic Effects

Axtmann [19] reports micro-seismic activity and a vertical displacement of < .4m/yr (4m total movement since 1956) at the Wairakei field which is an open loop, liquid-dominated resource without reinjection. On the other hand, Bowen [18] cites the rather remarkable lack of subsidence or seismic effects at the Larderello fields for the last 60 years and at The Geysers area over 12 years of operation. Both of these resources are vapor-dominated systems operated also as open systems without reinjection. Bowen [18] explains the differences in subsidence and seismic behavior between these systems by considering the physical characteristics of each type of reservoir. Rock formations in dry steam systems are usually competent, self-supporting structures, whereas those in hot water reservoirs might need internal fluid pressure to keep formations stable. Reinjection might be one technique of reducing subsidence and possibly enhancing reservoir lifetime.

Dry hot rock systems will probably have a closed, circulating loop with fluid passing through a fractured region of hot rock to remove heat. The current concept being considered by the Los Alamos Scientific Laboratory [8,10] will employ a two-hole. closed, pressurized water circulating system with a propped, vertical hydraulic fracture as the reservoir. Neither seismic nor subsidence problems are anticipated.

2.3.4 Gaseous Effluents

In any open circulating system without complete reinjection of geothermal fluid, there exists the potential for evolution of gases other than steam into the environment. Natural hydrothermal systems, both liquid- and vapor-dominated, have non-condensable gases dissolved in varying amounts in the geothermal fluid depending on the aquifer geochemistry. Experience at The Geysers, for instance, has shown that about 0.5% of the dry steam contains carbon dioxide CO_2 (0.4%), methane CH_4, hydrogen H_2, nitrogen N_2, ammonia NH_3, and hydrogen sulfide H_2S.[18] Similar effluents from Wairakei geofluids are cited by Axtmann. [19] Of these gases, H_2S and NH_3 pose the most serious health problem [17,18,19] but at present production rates at The Geysers and Wairakei these effluents,particularly H_2S levels, have not had significant environmental impact. However, even if a problem did exist, chemical treatment techniques to remove these effluents are under development.[18]

2.3.5 Liquid Effluents

Systems that discard spent geothermal liquids by means other than reinjection can also result in detrimental effects on the environment. Again, fluid composition is a function of the geochemistry of the aquifer and is obviously very site specific. For example, the highly saline geothermal brines of the Imperial Valley pose a potential hazard if they could enter the extensive agricultural irrigation system in that area.[18] In contrast, many natural geofluids are sufficiently pure to be used directly for domestic or industrial needs.[18] In some cases, poisonous materials such as arsenic, mercury and hydrogen sulfide are present in the liquid phase and constitute a hazard. Axtmann [19] cites evidence of possible food chain accumulation of mercury by organisms in the Waikato River, presumably caused by mercury contained in the water discharged at the Wairakei plant. In some cases where significant precipitation or scaling of solids, such as silica or calcium carbonate, occurs, the environmental effects of disposing of these materials might be important.

2.3.6 Thermal Discharges

The low efficiencies of geothermal generating plants result in a substantial amount of heat rejected to the environment for a specified power output. For instance, a 100 MW(e) plant using 200°C geothermal water, will reject ~550 MW as heat. However, because of this low efficiency, unit generating sizes are much smaller than many fossil-fuel or nuclear units [100 MW(e) or less versus 600-1500 MW(e)], and the total thermal discharge may be better distributed over a wider area. In addition expanded non-electrical uses of this lower grade heat could greatly reduce the impact of localized thermal discharges.

If the geothermal heat is rejected to an estuary or larger body of water, there are potential hazards to the aquatic ecology. An extreme example of this occurs at Wairakei [19] where water at temperatures as high as 60°C enters the Waikato River. In contrast, heat is rejected to the atmosphere at The Geysers using wet cooling towers,[8] but this requires significant quantities of water for evaporation. In areas where water supplies are limited, such as in Southwestern U.S., dry cooling towers or air-cooled condensers could be employed for atmospheric heat rejection.

For liquid-dominated or dry hot rock systems a closed, circulating loop has advantages over an open steam flashing system. Partially cooled geothermal liquid exiting from a heat exchanger or a flashing stage can be reinjected to the reservoir and thus reduce a fraction of the heat that would be rejected from an open system and presumably might even aid in sustaining reservoir lifetime.

In summary, the environmental impact of geothermal systems should be examined as an alternative to the impact of other sources of power. Geothermal energy development will be subject to a similar set of environmental regulations as are fossil-fuel and nuclear facilities.[17,20] Like any industry in its infancy, the geothermal community must be alert to environmental problems, and by this early awareness can approach reasonable solutions.

3. THERMODYNAMIC CRITERIA

3.1 Cycle Efficiencies

The maximum available energy produced as work (e.g., electrical) from any heat source is specified by the second law of thermodynamics. Its magnitude may be calculated by conducting a thought experiment in which heat (δQ_{gf}) is transferred from a high-temperature reservoir at T_{gf} (geothermal fluid) to a reversible Carnot heat engine that produces work (δW) in the surroundings and rejects heat (δQ_c) to a low-temperature reservoir at T_o (atmosphere or cooling water). The efficiency of this process is given by

$$\eta_c = \frac{\delta W}{\delta Q_{gf}} = \frac{T_{gf} - T_o}{T_{gf}} . \tag{1}$$

Because the rate at which fluid can be extracted from a geothermal liquid-dominated heat source is finite and because the liquid carries sensible, rather than latent, heat, it decreases in temperature as it transfers heat to the Carnot heat engine. Thus, the overall process must be envisioned as a summation over an infinite number of infinitesimally small engines each receiving a quantity of heat δQ^i_{gf} from a section of the reservoir at T^i_{gf} and rejecting a quantity δQ^i_c to a reservoir at T_o (assumed constant). The net maximum useful work for a geothermal liquid source at temperatures varying between T^{in}_{gf} and T^{out}_{gf} and a heat sink at ambient temperature T_o can be expressed as*

$$W^{max}_{net} = -\int_{T^{in}_{gf}}^{T^{out}_{gf}} \left[\frac{T^i_{gf} - T_o}{T^i_{gf}}\right] \delta Q^i_{gf} . \tag{2}$$

The idealized heat transfer process can be approximated by a constant pressure process and thus δQ^i_{gf} can be related to an enthalpy change of the fluid:

$$\delta Q^i_{gf} = dH\Big]_P = C_p dT^i_{gf} \text{ (per unit mass)}. \tag{3}$$

---

*One could include pressure energy but it usually represents only a few percent of the total available energy and is neglected in subsequent discussions.

By substituting Eq. (3) into Eq. (2), an equivalent expression for W_{net}^{max} results

$$W_{net}^{max} = -\int_{T_{gf}^{in}}^{T_{gf}^{out}} C_p dT_{gf}^i + T_o \int_{T_{gf}^{in}}^{T_{gf}^{out}} \frac{C_p dT_{gf}^i}{T_{gf}^i} . \qquad (4)$$

For the isobaric process (with no phase change) specified, the first integral of Eq. (4) is

$$\Delta H \Big]_{T_{gf}^{in}}^{T_{gf}^{out}} = \int_{T_{gf}^{in}}^{T_{gf}^{out}} C_p dT_{gf}^i ,$$

the second integral is

$$T_o \Delta S \Big]_{T_{gf}^{in}}^{T_{gf}^{out}} = T_o \int_{T_{gf}^{in}}^{T_{gf}^{out}} C_p \frac{dT_{gf}^i}{T_{gf}^i} ,$$

and Eq. (4) becomes

$$W_{net}^{max} = -\left(\Delta H - T_o \Delta S\right) \Big]_{T_{gf}^{in}}^{T_{gf}^{out}} = -\Delta B , \qquad (5)$$

which is commonly referred to as the change in availability, ΔB.

For a given fluid and a fixed T_o and reinjection temperature T_{gf}^{out}, which has a minimum value of T_o, the maximum work (per unit weight of geothermal fluid or per unit heat transferred) possible from an ideal, isobaric reversible process is a function of only T_{gf}^{in}, the geothermal source temperature. A plot of this value is given in Fig. 2. Any real process will have inefficiencies or nonreversible steps that will result in a net work less than W_{net}^{max}. These irreversibilities can be evaluated for each step of the conversion process. For example, neither the turbine nor the pump operates at 100% efficiency and frictional losses reduce the efficiency further. Any heat exchange step destroys availability. This step results in finite irreversibilities due to the finite temperature difference between each fluid necessary to transfer heat. Ideally, one could obtain more work from this process by operating a series of Carnot heat engines across any finite temperature difference present in the conversion process. A temperature-heat transferred or T-Q diagram* can be used to illustrate this concept for a heat exchanger.[1,4] Figure 3 is a T-Q diagram of heat exchange between a 280°C pressurized water geothermal fluid and refrigerant −11 (CCl_3F) used as the working fluid in a Rankine cycle. Optimal operation with the minimum irreversibility occurs when the heat capacities of both streams are constant. This situation produces a balanced exchanger in which the temperature difference ΔT between streams can be kept uniform. For the refrigerant shown in Fig. 3, the heat capacity C_p is far from constant at 48.3 bars (700 psia) between 50 and 218°C. This is the case for most of the hydrocarbons and halogenated hydrocarbons suitable for low-temperature power production. As the operating pressure is increased above the critical pressure of 44 bars (640 psia), to 69 bars (1000 psia), C_p becomes more uniform and ideal operation is approached. Thus, the critical properties of alternate working fluids relative to the geothermal source temperature are important in the initial screening process.

The overall heat exchange surface area A can be expressed as a function of the total heat Q, a mean $\overline{\Delta T}$, and an effective overall heat transfer coefficient U:

$$A = \frac{Q}{U\overline{\Delta T}} \tag{6}$$

A minimum practical $\overline{\Delta T}$ should be selected to keep the surface area and therefore cost at a reasonable level. In some cases the exchanger would reach a "pinched" condition whereby ΔT would approach zero at one or more locations. Then, a minimum ΔT would have to be specified for any potential pinch points. These concepts are applied to a number of specific cases in the discussions of power cycle thermodynamics and economics.

3.2 Thermodynamic Properties Evaluation

Selection of suitable secondary fluids and specification of optimum operating conditions are critically dependent on thermodynamic properties. Pressure-volume-temperature (PvT) data, heat capacity, and enthalpies of pertinent phase transformations are required for accurate

---

*Other diagrams such as temperature-entropy can be used to illustrate similar concepts.

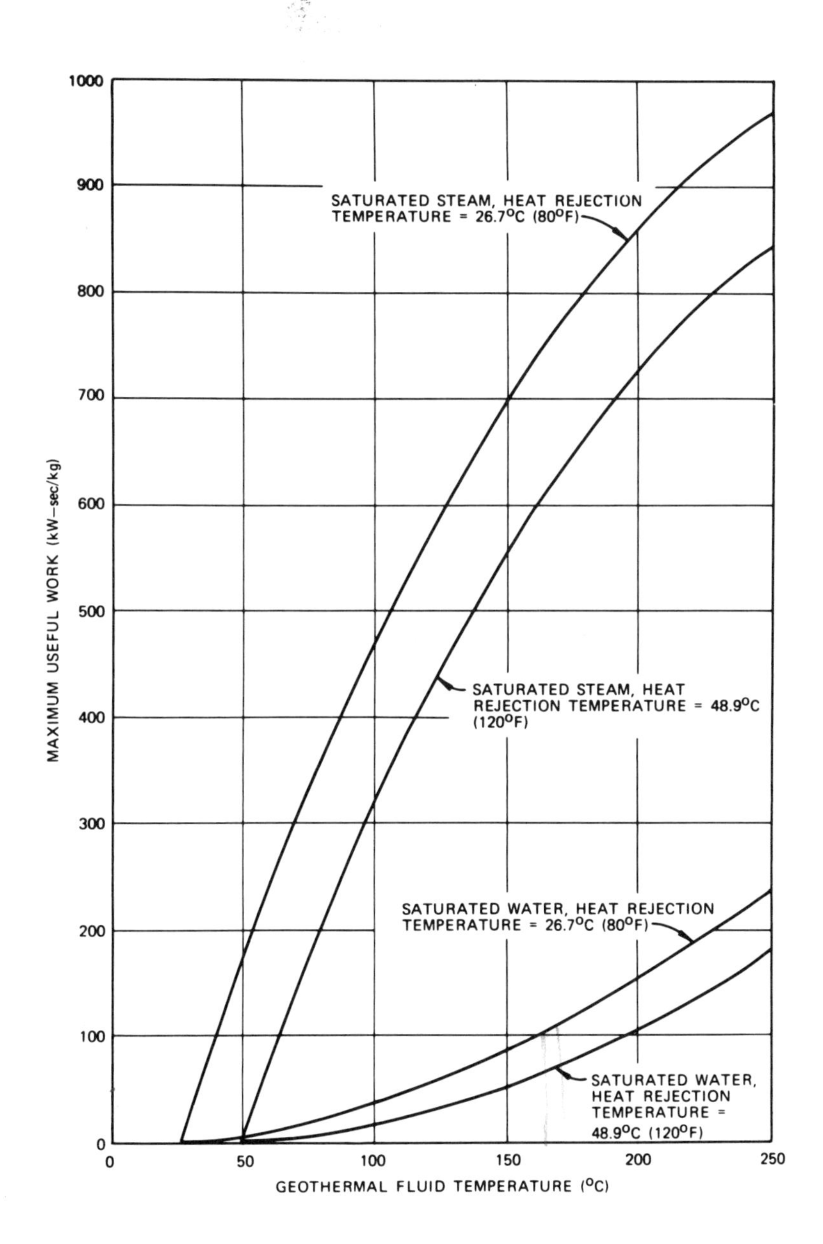

Fig. 2
Maximum useful work or availability (ΔB) plotted as a function of geothermal fluid temperature for saturated steam and saturated water sources.

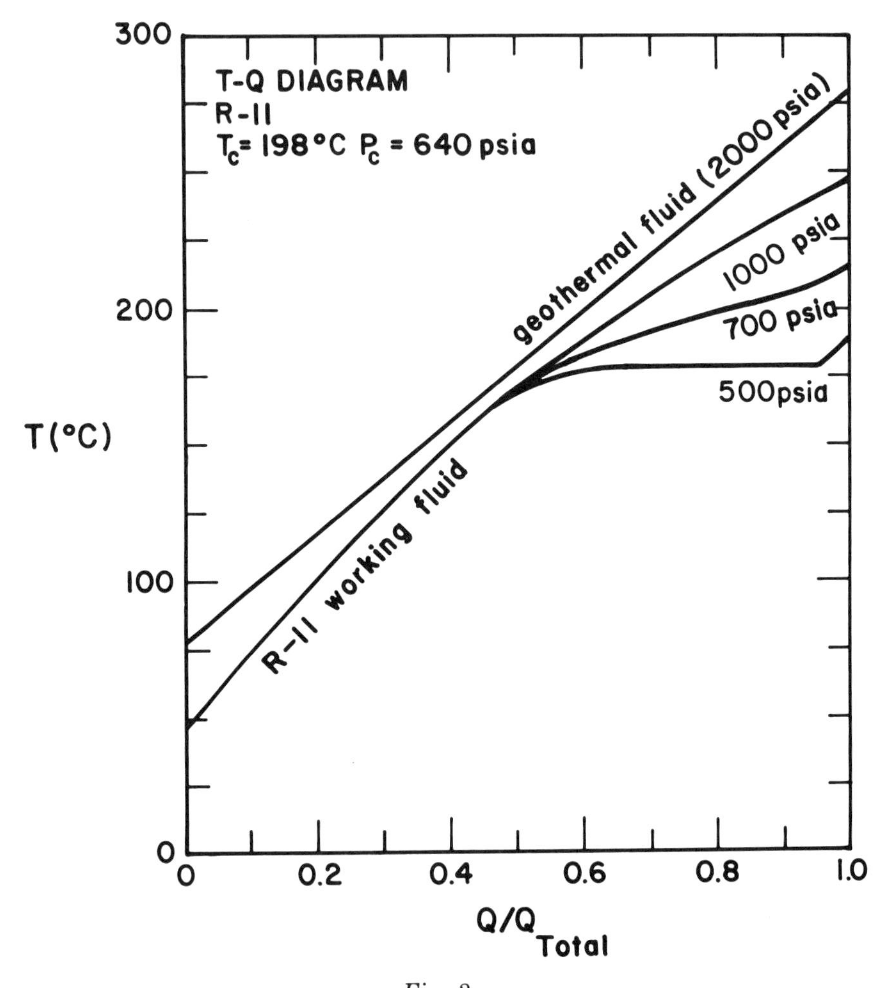

Fig. 3
Temperature-Heat Transferred (T-Q) diagram for R-11 (CCl_3F) with a 280°C compressed water geothermal source. Three operating pressures (500, 700, and 1000 psia) are shown.

cycle calculations. In addition, derived relationships of single phase entropy and enthalpy are also necessary. Two options can be pursued for arranging property values in suitable form: (1) available experimental data can be collected and suitable numerical or graphical interpolation and extrapolation procedures developed; or (2) an equation of state representing available PvT data can be used to generate any necessary properties when used in conjunction with a suitable expression for the infinite-dilution, (as $P \rightarrow 0$), ideal-gas-state heat capacity C_p^* over temperatures of interest. An equation of state is preferred when many iterative calculations requiring accuracy are performed.

A revised Martin-Hou equation of state [21,22] was selected for use in our cycle calculations because of a superior fit to experimental data in the supercritical region. (See Appendix E) In addition coefficients for the Martin-Hou equation for a number of refrigerants were available in the literature. A detailed description of the procedure used for parameter selection is described in an earlier paper by Milora,[21] and only the salient features are summarized here. The pressure is expressed as a function of the reduced temperature $T_r = T/T_c$, and volume, $v_r = v/v_o$ in a modified virial equation format as

$$P = \frac{P_c}{Z_c}\left[\frac{T_r}{\left(v_r - b\right)} + \frac{f_1\left(T_r\right)}{\left(v_r - b\right)^2} + \frac{f_2\left(T_r\right)}{\left(v_r - b\right)^3} + \frac{f_3\left(T_r\right)}{\left(v_r - b\right)^4} + \frac{f_4\left(T_r\right)}{\left(v_r - b\right)^5} + f_5\left(T_r\right)e^{-av_r} + f_6\left(T_r\right)e^{-2av_r}\right]. \quad (7)$$

Each dimensionless temperature-dependent coefficient $f_i(T_r)$ is empirically represented in the form

$$f_i\left(T_r\right) = A_i + B_iT_r + C_ie^{-KT_r} \qquad i = 1,\ldots,6 . \quad (8)$$

The first two functions, $f_1(T_r)$ and $f_2(T_r)$, can be expressed as explicit functions of the second and third virial coefficients, the critical volume v_c, and the empirical parameter b. The parameters a and b and the six functions $f_i(T_r)$ are chosen to fit experimental data along the critical isotherm $(T_r=1)$ including the critical point where

$$P = P_c \text{ and } \left.\frac{\partial P}{\partial v}\right)_{T_c} = \left.\frac{\partial^2 P}{\partial v^2}\right)_{T_c} = 0 .$$

Of a total of 21 constants, eight can be specified by fitting data along the critical isotherm. The six reduced virial coefficients $f_i(T_r)$ of Eq. (8) are alternatively expressed as functions of $f_i(T_r=1)$, B_i, C_i, and $K(i=1,6)$ by eliminating A_i from Eq. (8) using the critical isotherm $(T_r=1)$ behavior. Equation (8) can be rewritten as

$$f_i(T_r) = f_i(1) + B_i(T_r - 1) + C_i\left(e^{-KT_r} - e^{-K}\right). \qquad (9)$$

The remaining 13 constants are determined by solving a set of linear equations with an appropriate value for the nonlinear term K [Eq. (8)]. If available experimental data in various forms are used to specify the linear equations, an iterative procedure is followed with small adjustments to the parameters until agreement with experimental data is satisfactory. Alternatively, a nonlinear least squares technique could be used. These methods are empirical but ultimately provide a useful PvT equation of state for generating thermodynamic information (to accuracies better than 1%).

If experimental data are limited or nonexistent for a particular compound, an approximate set of initial parameter values is specified by utilizing generalized or corresponding states correlations. Because the corresponding states principle will be used extensively in the discussions that follow, a brief discussion of its theoretical basis is included here. PvT data from a wide variety of substances indicate a qualitatively similar behavior that can be correlated by a generalized relationship for the compressibility factor Z (Meissner and Seferian[23]);

$$Z = \frac{Pv}{RT} = f(T_r, P_r, Z_c) . \qquad (10)$$

Lydersen, Greenkorn, and Hougen [24] applied this concept to many PvT related properties, including the isothermal pressure dependence of enthalpy, entropy, and heat capacity. In addition, correlations have been established for thermodynamic properties at liquid-vapor equilibrium or saturation conditions including latent heat and saturated liquid and vapor densities.

Both enthalpy H and entropy S are thermodynamic state functions and therefore ΔH and ΔS are independent of the path selected to evaluate them. The ideal or infinitely-dilute gaseous heat capacity C_p^* is a function only of the temperature and any change in enthalpy along an isotherm is a function only of PvT properties. Consequently, ΔH or ΔS between state $\{T_1,P_1,v_1\}$ and state $\{T_2,P_2,v_2\}$ can be evaluated by a two-step process: (1) constant temperature $T = T_1$ from v_1 to v_2 or P_1 to P_2, and (2) constant volume $v \to \infty (P \to 0)$ path from T_1 to T_2. By introducing the reduced quantities $H_r = H/RT_c$ and $S_r = S/R$ as well as Z_c, T_r, P_r, and v_r, equations for ΔH and ΔS can be written as

$$\Delta H_r = \frac{\Delta H}{RT_c} = \frac{H(P_2, v_2, T_2)}{RT_c} - \frac{H(P_1, v_1, T_1)}{RT_c}, \qquad (11)$$

$$\Delta S_r = \frac{\Delta S}{R} = \frac{S(P_2, v_2, T_2)}{R} - \frac{S(P_1, v_1, T_1)}{R},$$

$$\Delta H_r = Z_c\left(P_{r2}v_{r2} - P_{r1}v_{r1}\right) + \left(T_{r1} - T_{r2}\right) + Z_c \int_{v_{r1}}^{\infty} \left[T_r \left(\frac{\partial P_r}{\partial T_r}\right)_{v_r} - P_r \right]_{T_r = T_{r1}} dv_r - Z_c \int_{v_{r2}}^{\infty} \left[T_r \left(\frac{\partial P_r}{\partial T_r}\right)_{v_r} - P_r \right]_{T_r = T_{r2}} dv_r + \int_{T_{r1}}^{T_{r2}} \frac{C_p^*}{R}\, dT_r \tag{12}$$

$$\Delta S_r = Z_c \int_{v_{r1}}^{\infty} \left[\left(\frac{\partial P_r}{\partial T_r}\right)_{v_r} \right]_{T_r = T_{r1}} dv_r - Z_c \int_{v_{r2}}^{\infty} \left[\left(\frac{\partial P_r}{\partial T_r}\right)_{v_r} \right]_{T_r = T_{r2}} dv_r + \int_{T_{r1}}^{T_{r2}} \frac{C_p^*}{RT_r}\, dT_r - \ell n T_{r2}/T_{r1} \ . \tag{13}$$

C_p^*/R is the reduced ideal gas heat capacity and is expressed in a power series expansion of reduced temperature as

$$C_p^*/R = C_o^* + C_1^* T_r + C_2^* T_r^2 + C_3^* T_r^3 \ . \tag{14}$$

The format of Eqs. (12) and (13) is convenient when using the Martin-Hou equation of state [Eq. (7)] because P_r and $(\partial P_r/\partial T_r)v_r$ can be evaluated and integrated term by term.[21] Other equivalent forms to Eqs. (12) and (13) might be used if an equation of state of different format was selected (see Reid and Sherwood[25] and Hougen, Watson, and Ragatz[26]).

In addition to the homogeneous gas or supercritical phase thermodynamic properties, saturation conditions also need to be specified. Experimental equilibrium vapor pressure (P^{sat}) data were fit to the following empirical equation using multiple linear regression,

$$\ell n P_r^{sat} = A_r - B_r/T_r - C_r \ell n T_r - D_r T_r + E_r T_r^2 \ , \tag{15}$$

with forced agreement at the critical point ($T_r = 1$, $P_r = 1$). Equation (15) is a smoothed function for the data and will provide a realistic value for the first derivative of P^{sat}.

The latent heat of vaporization ΔH_{vap} can then be estimated using the Clapeyron equation in reduced form:

$$\frac{\Delta H_{vap}}{RT_c} \equiv \frac{h_{fg}}{RT_c} = Z_c T_r \left(v_g - v_\ell\right)_r^{sat} \frac{dP_r^{sat}}{dT_r} \quad . \tag{16}$$

Gas and liquid phase specific volumes at saturation also are required. Since the Martin-Hou equation is not explicit in v_r, an iterative search procedure was used to calculate v_r^{sat} from known values of P_r^{sat} and T_r^{sat}. The liquid phase density $\rho_\ell^{sat} = 1/v_\ell^{sat}$ was expressed as a function of only T_r as

$$\rho_\ell)_r^{sat} = \frac{\rho_\ell^{sat}}{\rho_c} = 1 + A\left(1 - T_r\right)^{1/3} + B\left(1 - T_r\right)^{2/3} + C\left(1 - T_r\right) + D\left(1 - T_r\right)^{4/3} \quad . \tag{17}$$

The techniques described above provide the theoretical basis for predicting thermodynamic properties of working fluids considered in this paper (see Appendix E for details). Other procedures may be constructed which under certain circumstances might represent fluid properties just as accurately.

3.3 Working Fluid Choice

Seven secondary working fluids were examined in detail. Refrigerants R-22, R-717(NH_3), R-114, R-115, R-600a (isobutane), RC-318, and R-32 were selected because they provided a range of critical temperatures and pressures as well as molecular weight and because their thermodynamic properties were available in the literature.[27-37] In addition several others were considered on a preliminary basis as being potentially useful for geothermal applications. These included R-11, R-13B1, R-744 (CO_2), R-290, and R-113. Table 2 lists the pertinent thermodynamic properties, including parameters for the Martin-Hou equation of state, reduced heat capacity, reduced liquid density, and reduced vapor pressure correlations. Because Martin-Hou parameters were not available for difluoromethane (R-32) the corresponding-states principle was applied. The reduced coefficients for R-717 (NH_3) were used for R-32 since both had $Z_c = 0.242$.

Other factors besides thermodynamic properties frequently determine working fluid choice. These include fluid stability, flammability, toxicity, materials compatibility (corrosion), and cost. For the fluids considered, only flammability was considerably different in comparing the aliphatic hydrocarbons such as propane and isobutane to the halogenated aliphatic hydrocarbons and gases such as carbon dioxide. Ammonia does present a potential problem with chemical dissociation into hydrogen and nitrogen. Below 300°C, the rate of dissociation is relatively slow and should not present an overwhelming design limitation.

TABLE 2A

WORKING FLUID PROPERTIES

Compound	Formula	*M* g/gmole	T_c K	T_c °F	P_c bar[a]	P_c psia	v_c cm^3/g	v_c ft^3/lb	Z_c
R-11 trichlorofluoromethane	CCl_3F	137.38	471.2	388.4	44.13	640.0	1.804	0.0289	0.279
R-22 chlorodifluoromethane	$CHClF_2$	86.48	369.2	204.8	49.77	721.9	1.9060	0.03053	0.267
R-32 difluoromethane	CH_2F_2	52.03	351.56	173.1	58.30	845.6	2.3277	0.03729	0.242
R-113 trichlorotrifluoroethane	$C_2Cl_3F_3$	187.39	487.3	417.4	34.40	498.9	1.734	0.02778	0.276
R-114 dichlorotetrafluoroethane	$C_2Cl_2F_4$	170.94	418.9	294.3	32.61	473.0	1.7198	0.02753	0.275
R-115 chloropentafluoroethane	C_2ClF_5	154.5	353.1	175.9	31.57	458.0	1.6310	0.02613	0.271
R-13B1 bromotrifluoromethane	$CBrF_3$	148.93	340.2	152.6	39.64	575.0	1.3426	0.02151	0.280
R-600a isobutane	C_4H_{10}	58.12	408.1	275.0	36.48	529.1	4.5220	0.07244	0.283
R-717 ammonia	NH_3	17.03	405.4	270.1	112.78	1635.7	4.2470	0.06803	0.242
RC-318 octafluorocyclobutane	C_4F_8	200.04	388.5	239.6	27.83	403.6	1.6130	0.02584	0.278
R-744 carbon dioxide	CO_2	44.01	304.2	87.9	73.77	1070.0	2.1372	0.03424	0.274
R-290 propane	C_3H_8	44.10	370.0	206.3	42.57	617.4	4.5437	0.07278	0.277
Water	H_2O	18.02	647.3	705.5	221.18	3208.0	3.1077	0.04978	0.230

[a] 1bar = 10^5 Pa

TABLE 2B

WORKING FLUID PROPERTIES

Compound	$C_p^*/R = C_o^* + C_1^* T_r + C_2^* T_r^2 + C_3^* T_r^3$				$\rho_\ell)_r^{sat} = 1 + A(1-T_r)^{1/3} + B(1-T_r)^{2/3} + C(1-T_r) + D(1-T_r)^{4/3}$			
	C_o^*	C_1^*	C_2^*	C_3^*	A	B	C	D
R-11 trichlorofluoromethane	---	---	---	---	---	---	---	---
R-22 chlorodifluoromethane	2.6073	5.6582	-0.5479	-0.1943	1.6677	1.1218	-0.6805	0.6249
R-32 difluoromethane	4.2778	3.2980	0	0	1.4083	3.3657	-5.0893	3.6366
R-113 trichlorotrifluoroethane	---	---	---	---	---	---	---	---
R-114 dichlorotetrafluoroethane	2.5063	22.6489	-8.1711	0	1.9282	-0.1467	1.4087	-0.5892
R-115 chloropentafluoroethane	2.4760	14.0108	-3.6790	0	2.4505	-3.3572	6.5815	-3.297
R-13B1 bromotrifluoromethane	---	---	---	---	---	---	---	---
R-600a isobutane	-0.9512	20.4097	-4.6066	0.4078	1.1730	4.6140	-7.5010	4.7699
R-717 ammonia	4.3032	-1.8411	3.1995	-0.9830	1.5489	1.9711	-1.5957	1.0322
RC-318 octafluorocyclobutane	3.2671	26.0428	-8.1154	0.7411	1.8310	0.6101	0.4132	-0.2306
R-744 carbon dioxide	---	---	---	---	---	---	---	---
R-290 propane	---	---	---	---	---	---	---	---
Water	3.5913	0.86004	0.00967	0	---	---	---	---

TABLE 2C

WORKING FLUID PROPERTIES

Compound	$\log P)_r^{sat} = A_r - B_r/T_r - C_r \ell n T_r - D_r T_r + E_r T_r^2$					References
	A_r	B_r	C_r	D_r	E_r	
R-11 trichlorofluoromethane	---	---	---	---	---	26,27
R-22 chlorodifluoromethane	20.5140	7.2003	-6.8419	19.4130	6.0993	27,28 $(C_p^*$ and $\rho_\ell)$
R-32 difluoromethane	15.3618	9.1356	0	10.4699	4.2410	27,29 (ρ_ℓ, P_r^{sat})
R-113 trichlorotrifluoroethane	---	---	---	---	---	27
R-114 dichlorotetrafluoroethane	38.4406	3.5155	-25.5945	47.9292	13.0041	27,30 (C_p^*), 31 (ρ_ℓ)
R-115 chloropentafluoroethane	4.4096	12.5996	13.9492	-8.1830	0	27,32 (ρ_ℓ), 33 (C_p^*)
R-13B1 bromotrifluoromethane	---	---	---	---	---	27
R-600a isobutane	10.8905	8.7656	2.6561	4.7799	2.6550	27,34,26 (C_p^*), this work $(\rho_\ell$ and $P_r^{sat})$
R-717 ammonia	7.3672	10.8839	8.5674	-2.4647	1.0520	27, this work $(C_p^*, \rho_\ell$, and $P_r^{sat})$
RC-318 octafluorocyclobutane	87.3214	-13.9947	-93.6712	130.3978	29.0818	27,35 $(C_p^*$ and $\rho_\ell)$, this work (P_r^{sat})
R-744 carbon dioxide	---	---	---	---	---	25,27
R-290 propane	---	---	---	---	---	27,25
Water	---	---	---	---	---	25,26 (C_p^*), 36 (P_r^{sat}) 37 (general properties)

TABLE 2D

MARTIN-HOU EQUATION OF STATE PARAMETERS

Compound		a	b	K	$f_1(1)$	$f_2(1)$	$f_3(1)$	$f_4(1)$	$f_5(1)$
R-22	$CHClF_2$	16.73381	0.065520	4.2	-1.356568	0.721958	-0.0366074	-0.0617381	9.325812×10^3
R-717 R-32	NH_3 CH_2F_2 [a]	14.33333	0.018333	6.42	-1.545953	1.022201	-0.1891483	-0.0458976	1.741376×10^3
R-114	$C_2Cl_2F_4$	16.733805	0.214829	3.0	-1.458334	1.065470	-0.390954	0.0577140	---
R-115	C_2ClF_5	16.733805	0.233815	5.48	-1.476552	1.092186	-0.405404	0.0653873	---
RC-318	C_4F_8	16.733805	0.232642	5.0	-1.442633	1.044207	-0.379554	0.0554769	---
R-600a	C_4H_{10}	16.024735	0.1492517	3.97	-1.345236	0.813206	-0.182000	-0.00369858	7.369969×10^3

Compound		$f_6(1)$	B_1	B_2	B_3	B_4	B_5	B_6
R-22	$CHClF_2$	---	0.635479	0.659672	-1.021552	0.497061	-4.11421×10^4	---
R-717 R-32	NH_3 CH_2F_2 [a]	1.803382×10^5	1.658608	-3.768077	4.924337	-1.994371	2.56467×10^4	-7.2990×10^6
R-114	$C_2Cl_2F_4$	---	0.625670	-0.112213	---	0.0173824	---	---
R-115	C_2ClF_5	---	0.831589	-0.362615	---	0.0556471	---	---
RC-318	C_4F_8	---	0.710469	-0.191577	---	0.0345776	---	---
R-600a	C_4H_{10}	4.1502651×10^6	0.5659241	0.19349381	-0.185239	0.0679934	-1.647185×10^4	1.437288×10^7

Compound		C_1	C_2	C_3	C_4	C_5	C_6	References
R-22	$CHClF_2$	-17.506396	19.310496	---	-2.577068	---	---	28
R-717 R-32	NH_3 CH_2F_2 [a]	-81.254057	-245.44040	647.520530	-323.02139	4.294690×10^6	-9.571300×10^8	21, this work
R-114	$C_2Cl_2F_4$	-5.043394	4.566885	---	-0.374187	---	---	30
R-115	C_2ClF_5	-26.570529	23.639653	---	-1.313044	---	---	32
RC-318	C_4F_8	-29.44543	26.503886	---	-2.301590	---	---	35
R-600a	C_4H_{10}	-12.09121	15.084080	-2.022455	-1.350682	2.009688×10^5	5.511819×10^8	this work

[a] Parameters for R-717 (NH_3) used for R-32.

4. POWER CYCLE THERMODYNAMICS

4.1. Available Energy of Low-Temperature Liquid and Steam Systems

As stated earlier, the effectiveness of any energy conversion system depends on the nature of the heat source as well as prevailing ambient heat rejection conditions. For example, an enormous amount of heat is contained in the earth's oceans; but in reality this heat is of little use, because its temperatures are very near the local ambient values. In some cases, lower temperature water is available at depths less than 1000 m suitable for heat rejection, but conversion efficiencies are still very low. This unavailability of low temperature heat is, of course, a consequence of the second law of thermodynamics which places quantitative limits on the maximum amount of useful work that can be obtained by operating a heat engine between any given heat source and sink combination.

In terms of the fluid properties enthalpy, H, and entropy, S, the potential maximum useful work is equivalent to the change in availability (ΔB) of the fluid between the conditions at the wellhead (P_{gf},T_{gf}) and the ambient or dead state (P_o,T_o).

$$W_{net}^{max} = \left(\Delta H - T_o \Delta S\right)\Bigg|_{T_o, P_o}^{T_{gf}, P_{gf}} \tag{18}$$

In many practical situations, the effect of pressure on the availability of the subcooled liquid is negligible and thus a simple expression for thermally-transferrable W_{net}^{max} can be written in terms of the liquid C_p (assumed constant), the temperatures T_{gf} and T_o and, in the case of saturated steam, h_{fg}:

Saturated Liquid:

$$W_{net}^{max} = C_p\left\{\left(T_{gf} - T_o\right) - T_o \ln\left(\frac{T_{gf}}{T_o}\right)\right\} . \tag{19}$$

Saturated Steam:

$$W_{net}^{max} = h_{gf}\left(1 - \frac{T_o}{T_{gf}}\right) + C_p\left\{\left(T_{gf} - T_o\right) - T_o \ln\left(\frac{T_{gf}}{T_o}\right)\right\} . \tag{20}$$

In terms of a specific physical situation, Eq. (19) describes a process in which an ideal heat engine extracts an amount of heat, $C_p(T_{gf}-T_o)$, from the saturated liquid until its temperature is reduced to ambient. Part of this heat is converted reversibly to work and the rest $C_pT_o\ell n\ T_{gf}/T_o$ is rejected to the sink as waste heat at temperature T_o. Pressure energy not removed from the source fluid may be used in part or entirely for reinjection.

Equation (20) can be thought of as representing two consecutive reversible processes; the first of which accepts heat (h_{fg}) from the condensing vapor at constant temperature T_{gf}, produces useful work, and rejects an amount of heat $h_{fg}T_o/T_{gf}$ to the atmosphere. The second process extracts the remainder of heat from the liquid condensate as described above. For the modest temperatures considered here, the former process contributes significantly more work than the latter, owing to large values of the ratio $h_{fg}/C_p(T_{gf} - T_o)$. This fact is demonstrated in Fig. 2. Below 200°C, the superiority of steam is particularly evident. This is important from the standpoint of power plant capital costs, since it is reasonable to assume that costs will depend strongly on the amount of geothermal fluid required to provide the desired generating capacity. In this respect, the cost of power derived from geothermal flashing plants could not *a priori* be considered comparable to the cost of power derived from vapor-dominated fields.

Figure 2 also illustrates the pronounced effect of heat rejection temperature on the potential of hot water systems, particularly below 200°C. Reducing the temperature at which waste heat is rejected improves resource utilization by an amount greater than a corresponding increase in reservoir temperature. If practical, it is clearly desirable to utilize lower heat rejection temperatures when and where environmental conditions permit. Unfortunately, because of economic limitations, even relatively high-temperature geothermal flashing plants could not be expected to take full advantage of very low ambient temperatures. This point is discussed further in the section on turbine size requirements.

4.2 Alternate Fluid Cycle Design Criteria

In reality, perfect heat engines are nonexistent and, at best, only a fraction of the theoretically available work could actually be obtained from any given reservoir/environment combination. If the cost of drilling wells is high and/or the reservoir lifetime is uncertain, then there is a stronger incentive to utilize the resource efficiently, particularly when dealing with low-temperature hot water systems. Summarizing the concepts introduced previously, efficient resource utilization will result when:

1. Most of the heat is extracted from the geothermal fluid before disposal or reinjection.

2. Temperature differentials across heat transfer surfaces are maintained at minimum practical levels.

3. Turbines and feed pumps are carefully designed for optimum efficiency.

4. Heat is rejected from the thermodynamic cycle at a temperature near the minimum ambient temperature.

Whether or not these conditions are met depends largely upon the choice of thermodynamic working medium, the geothermal fluid temperature, and the temperature of the coolant (water or air) to which the power plant rejects waste heat. For example, if waste heat is to be rejected from the thermodynamic cycle at a constant temperature by a condensing vapor (as in condition 4 above), then the compound's critical temperature must be greater than the temperature of the power plant coolant. As will be shown in Section 4.3, fulfillment of conditions 1 and 2 suggests the use of supercritical Rankine cycles. Thus, the critical temperature of the working fluid should also be below the maximum geothermal fluid temperature.

These requirements restrict the number of compounds that can be used to those that have critical temperatures between approximately 25 and 200°C. This temperature range is broad enough to qualify a variety of low boiling point compounds that have heretofore been considered primarily for use in refrigeration systems; namely, the lower hydrocarbons and their halogenated derivatives and ammonia. Our selection of seven representative compounds, whose thermodynamic properties are tabulated in Table 2 was consistent with these criteria. Although these compounds are typical of the kind being considered for various low temperature, Rankine-cycle applications, it should not be expected that individual compounds will perform alike (in a thermodynamic sense) because there exist significant differences in the two properties, the critical temperature T_c and the reduced, constant pressure specific heat of the fluid, that affect performance most. Indeed, we will demonstrate shortly that the thermodynamic performance varies considerably from compound to compound and that each compound performs best at a definite geothermal fluid temperature.

4.3 Single-Fluid Cycle Optimization

Quantitatively, one can define a cycle efficiency factor that is a measure of the fraction of the available energy of the geothermal fluid that is converted into useful work. A utilization factor, η_u, can be defined in terms of the net power, $\mathcal{P}$, the geothermal flow rate $\dot{m}_{gf}$, and the availability $-\Delta B = W_{net}^{max}$:

$$\eta_u = \frac{P}{\dot{m}_{gf} W_{net}^{max}} = \frac{P}{\dot{m}_{gf}\left[C_p\left(T_{gf} - T_o\right) - C_p T_o \ln\left(\frac{T_{gf}}{T_o}\right)\right]} . \quad (21)$$

η_u is a direct measure of the efficiency of resource utilization, because for a fixed T_{gf}, higher values of η_u correspond to lower well flow rates for a given power output. η_u should not be confused with the thermodynamic cycle efficiency, η_{cycle}, which is a measure of how efficiently the transferred geothermal heat is converted into work.

If, in this heat transfer process, fluid is cooled to some intermediate temperature T_{gf}^{out}, then η_{cycle} becomes:

$$\eta_{cycle} = \frac{P}{\dot{m}_{gf} C_p\left(T_{gf}^{in} - T_{gf}^{out}\right)} . \quad (22)$$

When little heat is being extracted from the geothermal fluid ($T_{gf}^{in} \cong T_{gf}^{out}$), η_{cycle} will be proportionally greater than η_u because the resource is being utilized poorly. This is true because the availability of the discarded or reinjected fluid at temperatures above ambient is not considered as a thermodynamic loss in calculating η_{cycle}.

In principle, η_u can assume any value between zero (P=0) and unity ($P = \dot{m}_{gf} W_{net}^{max}$); while in practice its value is determined from economic considerations by balancing the cost of obtaining the heat (drilling and piping costs) against the cost of processing it (heat exchangers, turbines, pumps) to generate electricity in the power station. If η_u is small, then the resource is being utilized poorly and a large investment in wells is required (cost per unit power $\rightarrow \infty$ as $\eta_u \rightarrow 0$). On the other hand, if we approach utilization of the full potential of the resource, then total well costs will be lower but the required investment in highly efficient power conversion equipment will be large (cost per unit power $\rightarrow \infty$ as $\eta_u \rightarrow 1$). The economic optimum occurs when η_u assumes some intermediate value,for example at The Geysers where a value of 0.55 is typical for T_o = 26.7°C (80°F). In the examples that follow, the emphasis is placed on maximizing η_u within reasonable limits since the relative cost of obtaining geothermal heat is expected to be high, particularly for low-temperature liquid-dominated resources. η_u will be expressed in terms of performance of major plant components, including heat exchangers, condensers, turbines, and pumps.

In the following analysis cycle performance is calculated using a set of equations which describe the work and heat flow rates to and from the major plant components (see Appendix E). The net power, P, of the plant was specified as was the heat rejection temperature T_o. Heat exchanger performance is also regulated by limiting minimum approach temperatures to 10°C with a 15°C difference between the outlet working fluid and inlet geothermal fluid temperature. A preliminary analysis indicates that these temperature differences are within the range for optimum economic performance (see Appendix B). Air or liquid cooled condenser conditions are also controlled in a similar manner to specify the condenser outlet working fluid temperature at T_o. After selecting a working fluid and a geothermal fluid inlet temperature, and fixing the above conditions, the major independent variable is the maximum cycle operating pressure P at the turbine inlet. The calculational details of cycle design are discussed in the following example.

Figure 4A depicts on a temperature-enthalpy diagram the key thermodynamic steps involved in Rankine cycle operation. The compound chosen to illustrate the process is R-115 (C_2ClF_5) which has a critical temperature of 80°C and is therefore a suitable working fluid for both subcritical and supercritical operation at a geothermal fluid temperature of 150°C.

To begin the process, saturated liquid is pumped from state point E (where it is at the vapor pressure corresponding to the minimum cycle temperature T_o) to the maximum cycle operating pressure of 27.5 bars ($P_r = 0.87$). The work (per unit weight) expended in this step is given by the difference in specific enthalpy of the fluid between state points A and E:

$$W_p \equiv H_A - H_E \simeq \frac{\bar{V}_\ell^{sat}\left[P_A - P_{sat}\left(T_o\right)\right]}{\eta_p} . \tag{23}$$

In all of the examples to be discussed, a value of 0.8 is assumed as a representative pump efficiency η_p.

The liquid, which is now subcooled at a temperature just slightly higher than T_o, is heated by the geothermal fluid until its temperature reaches the saturation value corresponding to the cycle operating pressure, in this case 73°C. Further heat addition causes the fluid to change phase at constant temperature until the liquid is completely vaporized. Superheat is then added to the vapor until the top cycle temperature corresponding to state point B is reached. This temperature was chosen to be 15°C below the maximum geothermal fluid temperature.

The straight line segment drawn above the working fluid heating path represents the temperature-enthalpy history of the geothermal fluid (liquid) as it passes through the heat exchanger. A "pinch" or minimum temperature difference occurs in the heat exchanger at the point where the liquid begins to evaporate. In all examples to be considered, a 10°C minimum temperature difference is assumed. The temperature and enthalpy of the working fluid at the pinch point then uniquely determine the temperature T_{gf}^{out} at which the geothermal fluid is discarded from the heat exchanger. In this example, the outlet temperature is small (47°C) and the residual availability of the discarded liquid is quite low (see Fig. 2). An energy balance for the entire heat addition process relates the geothermal fluid and working fluid mass flow rates to the changes in geothermal fluid temperature and specific enthalpy of the working fluid between state points A and B,

$$\dot{m}_{gf} C_{p_{gf}} \left(T_{gf} - T_{gf}^{out} \right) = \dot{m}_{wf} \left(H_B - H_A \right) \quad . \tag{24}$$

Owing to the constant temperature phase change of the working fluid, the mean heat exchanger temperature difference is large at this operating pressure. Although this difference favors the heat transfer process, thus requiring less heat transfer surface area, it will be shown that this results in a degradation of availability and a corresponding loss in thermodynamic performance of the process.

The next step in the cycle involves the turbine expansion from state point B to a lower pressure corresponding to the vapor pressure $P^{sat}(T_0)$ of the fluid at the temperature at which heat is rejected in the condenser. The enthalpy of the vapor at the turbine exhaust (state point C) is determined by the net turbine efficiency η_t and the enthalpy change that the vapor would experience if it were to expand adiabatically (no heat losses) with negligible friction losses to the exhaust pressure $P^{sat}(T_0)$. This ideal path, which is shown terminating at point C′, represents a constant entropy expansion from state point B to the specified pressure. The specific work developed by the turbine during an actual expansion is expressed in terms of this isentropic enthalpy drop by

$$W_T \equiv H_B - H_C = \eta_t \left\{ H_B \left[S_B, P_B \right] - H_{C'} \left[S_B, P^{sat} \left(T_0 \right) \right] \right\} \quad . \tag{25}$$

Turbine stage efficiencies, η_t, of 0.85 are typical, unless the turbine expansion path enters the vapor-liquid equilibrium dome. When condensation occurs within a turbine stage, the efficiency of that stage deteriorates due largely to impingement of condensate on the turbine blade surfaces. Accordingly, for a small amount of moisture, wet stage efficiency is usually written as the difference between dry stage efficiency and the mass fraction of liquid χ,

$$\eta_t \simeq \eta_{t_{dry}} - \chi$$

with

$$\eta_{t_{dry}} = 0.85 \quad . \tag{26}$$

At the turbine exhaust, the working fluid may still be superheated above the minimum heat rejection temperature. This superheat, which must be removed from the vapor before condensation can proceed, represents a thermodynamic loss since additional work could still, in principle, be obtained from the hot vapor as it cools at constant pressure to the saturation point (state point D). In Fig. 4A, a considerable amount of superheat is shown by the large enthalpy difference between state points C and D. Large amounts of superheat also increase the required surface area of heat rejection equipment, since practical gas-phase heat transfer coefficients are much lower than condensing coefficients. An alternative would be to use direct contact heat exchange.

After desuperheating takes place, the remaining waste heat is removed in the condenser at constant temperature (T_0) and pressure [$P^{sat}(T_0)$] as the fluid changes phase from saturated vapor to saturated liquid. A condensing temperature of 26.7°C (80°F) was assumed.

The total sensible and latent heat removed from the cycle is given by the difference in enthalpy between state points C and E,

$$Q_{rej} = H_C - H_E \quad . \tag{27}$$

The net power developed by the cycle is the difference between the turbine output and the pumping power required to raise the pressure of the condensate to the cycle operating pressure. Secondary parasitic losses including loop pressure drops, heat losses, and auxiliary power in excess of primary pumping power, are neglected. For a given power rating, P, this specifies the working fluid flow rate and hence, by Eq. (24), the geothermal fluid flow:

$$P = \dot{m}_{wf}\left\{\left(H_B - H_C\right) - \left(H_A - H_E\right)\right\} . \tag{28}$$

Alternatively, the power can be expressed as the difference between the rate of heat addition in the preheater/evaporator/superheater and the rate of heat rejection from the desuperheater and condenser as

$$P = \dot{m}_{wf}\left\{\left(H_B - H_A\right) - \left(H_C - H_E\right)\right\} . \tag{29}$$

The cycle and utilization efficiencies can now be expressed solely in terms of the cycle operating conditions and the temperatures of the geothermal fluid before and after the heat transfer process, assuming constant geothermal fluid heat capacity.

$$\eta_{cycle} = 1 - \frac{\left(H_C - H_E\right)}{\left(H_B - H_A\right)} , \tag{30}$$

$$\eta_u = \frac{\left(T_{gf} - T_{gf}^{out}\right) \quad \eta_{cycle}}{\left\{T_{gf} - T_o\right\} - T_o \ln\left(\frac{T_{gf}}{T_o}\right)} \quad . \tag{31}$$

Although the geothermal fluid is cooled to 47°C in this example, the heat is being utilized poorly because of the large amount of sensible heat that is rejected at high temperatures in the desuperheater. The resulting low value of cycle efficiency (9%) is therefore responsible for the poor utilization efficiency (47%). Equation (31) expresses the need for simultaneous high cycle efficiency and a large amount of heat extraction from the geothermal fluid.

To improve performance, the cycle operating pressure is increased in Fig. 4B to 39.26 bars (P_r = 1.24). At supercritical pressures, the constant temperature phase change shown in Fig. 4A does not occur and temperature differences in the heat exchanger are reduced. This more efficient heat addition process improves the thermodynamic cycle efficiency by moving the turbine inlet and outlet enthalpies (state points B and C, respectively) closer to the vapor equilibrium line. Accordingly, less heat is rejected in the desuperheater and the cycle efficiency is increased to 11.2%. Since the heat exchanger outlet temperature is essentially unchanged from that of the previous example, the utilization efficiency increases proportionately to 56.5%.

In Fig. 4C, the operating pressure is increased further to 80.1 bars (P_r = 2.54). The heat transfer path at this pressure is almost ideal in the sense that a nearly uniform temperature difference between the two fluids is maintained throughout the entire heat exchange process. This uniformity is attributable to the constancy of specific heat of the working fluid at higher reduced pressures. The turbine expansion path extends very near the vapor saturation line and only a small amount of superheat must be removed from the exhaust gas to reduce its temperature to the saturation value. The combined effects of high cycle efficiency (11.9%) and a high degree of heat extraction from the geothermal fluid result in a high value of utilization efficiency (63.2%).

To eliminate all superheat at the turbine exhaust, it is necessary to increase the cycle pressure even further as in Fig. 4D where the operating pressure is 114.4 bars(P_r = 3.62). The efficiency of the heat transfer process is essentially unchanged from the previous example; however, the turbine expansion path, which has been displaced further to the left, terminates just inside the two-phase region with a moisture content of about 2%. As much as 5% moisture is present in the latter part of the turbine expansion with the result that net turbine efficiency drops below the nominal dry stage value of 85%. This in turn adversely affects the cycle efficiency which, in this case, has dropped to 10.6%. At this high

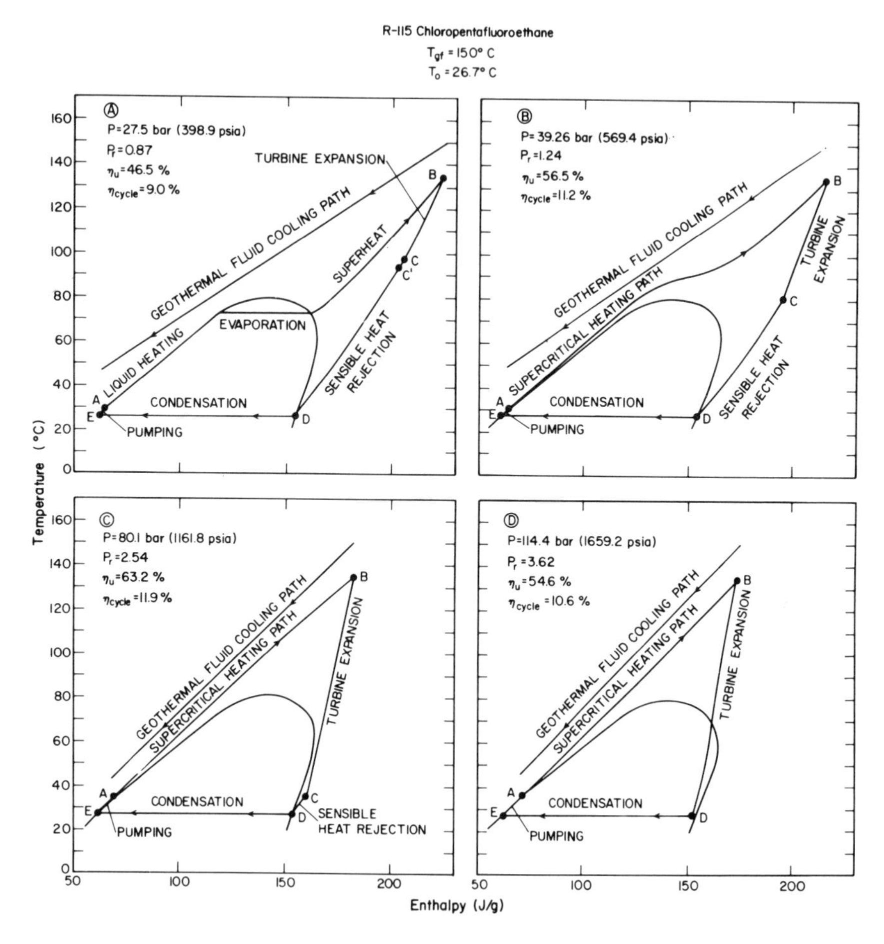

Fig. 4
Approach to the thermodynamically optimized Rankine cycle for R-115 with a 150°C liquid geothermal fluid source and heat rejection at 26.7°C. Temperature-Enthalpy diagrams shown at different reduced cycle pressures.

operating pressure, the "boiler feed" pump work has increased substantially as measured by the difference in enthalpy between state points A and E. The large pumping power requirement, which amounts to nearly 50% of the net turbine power, preheats the working fluid to 36°C before it enters the heat exchanger. Because of the assumed pinch point temperature difference of 10°C, the minimum temperature to which the geothermal fluid can be cooled is 46°C. Since the amount of heat extraction and cycle efficiency have both decreased from the previous example, it follows that the utilization factor must also drop to a lower value (54.6%). Obviously, any further increase in operating pressure will result in even poorer performance as the pump and turbine efficiencies worsen.

4.4 Irreversibility Analysis of Thermodynamic Performance

Quantitatively, the effect of operating pressure on overall system performance can best be explained by considering the performance of individual cycle components. The concept of thermodynamic irreversibility is useful for identifying where major system inefficiencies exist. For any process in which a material undergoes a change of thermodynamic state, the value of the irreversibility function I is a measure of the amount of potential work which is lost from the system due to such irreversible processes as friction in turbines and pumps and finite temperature differences incurred across heat exchange surfaces. For a particular process involving a change in state of the working fluid, the value of this irreversibility is given by the product of the minimum ambient temperature and the net amount of entropy created in a system which comprises the working fluid process and those parts of the surroundings such as heat reservoirs and the atmosphere which exchange heat with it.

Four major steps of the thermodynamic cycle are potential sources of irreversibility: (1) pumping of the condensate to operating pressure, (2) transfer of heat between the geothermal fluid and the working fluid, (3) the turbine expansion path, and (4) the rejection of sensible heat to the atmosphere in the desuperheater. In addition, the residual availability of the geothermal fluid rejected from the heat exchanger is an important loss which must be considered.

Referring to the example of Fig. 4, the following expressions relate individual component irreversibilities (per unit mass of working fluid) to the thermodynamic cycle conditions:

Feed pump irreversibility:

$$I = \left(1 - \eta_p\right)\left(H_A - H_E\right) \simeq \frac{\left(1 - \eta_p\right)}{\eta_p} v_\ell^{sat}\left(T_o\right)\left[P_A - P^{sat}\left(T_o\right)\right] \tag{32}$$

Primary heat exchanger irreversibility:

$$I = \frac{T_o\left(H_B - H_A\right)}{\left(T_{gf}^{in} - T_{gf}^{out}\right)} \int_{T_{gf}^{out}}^{T_{gf}^{in}} \frac{T_{gf} - T_{wf}}{T_{gf}T_{wf}}\, dT_{gf} \; . \tag{33}$$

Turbine irreversibility:

$$I = T_o \int_{H_{C'}}^{H_C} \frac{dH}{T} \simeq \frac{T_o}{T_{ex}} \left[\frac{1 - \eta_t}{\eta_t} \right] \left(H_B - H_C \right) \quad . \qquad (34)$$

Sensible heat rejection (desuperheat) irreversibility:

$$I = \left(H_C - H_D \right) - T_o \int_{H_D}^{H_C} \frac{dH}{T} \quad . \qquad (35)$$

Residual availability or irreversibility of reinjected geothermal fluid:

$$I = \frac{\left(H_B - H_A \right)}{\left(T_{gf}^{in} - T_{gf}^{out} \right)} \left\{ T_{gf}^{out} - T_o - T_o \ln \left(\frac{T_{gf}^{out}}{T_o} \right) \right\} \quad . \qquad (36)$$

Equations (32) through (36) indicate the need to maintain high pump and turbine efficiencies while minimizing such quantities as heat exchanger temperature differences, superheat at the turbine exhaust, and the terminal temperature of the geothermal fluid in the primary heat exchanger. At best, these conditions can be fulfilled only partially at any specified operating pressure, and consequently optimum performance occurs at some compromise pressure where the total, or sum, of all irreversibilities is a minimum. This fact is illustrated in Fig. 5 where the pertinent irreversibility functions are plotted against reduced operating pressure for the examples of Fig. 4. At subcritical ($P_r<1$) and just slightly supercritical ($P_r>1$) operating pressures, the "thermal" inefficiencies created by large temperature differentials in the primary heat exchanger and high turbine exhaust and residual fluid temperatures are large compared to the mechanical inefficiencies of the turbine and feed pump. The heat exchanger irreversibility dominates and is primarily responsible for the correspondingly low values of η_u.

As the pressure is increased, both pump and turbine output ratings increase and, hence, their irreversibilities increase steadily. At intermediate pressures, this disadvantage is more than offset by the improvement in primary heat exchanger and desuperheater performance. Accordingly, the utilization factor increases. This trend continues up to $P_r \cong 2.5$. At that point the heat exchanger is approaching the ideal behavior indicated in Fig. 4C by the nearly linear temperature-enthalpy

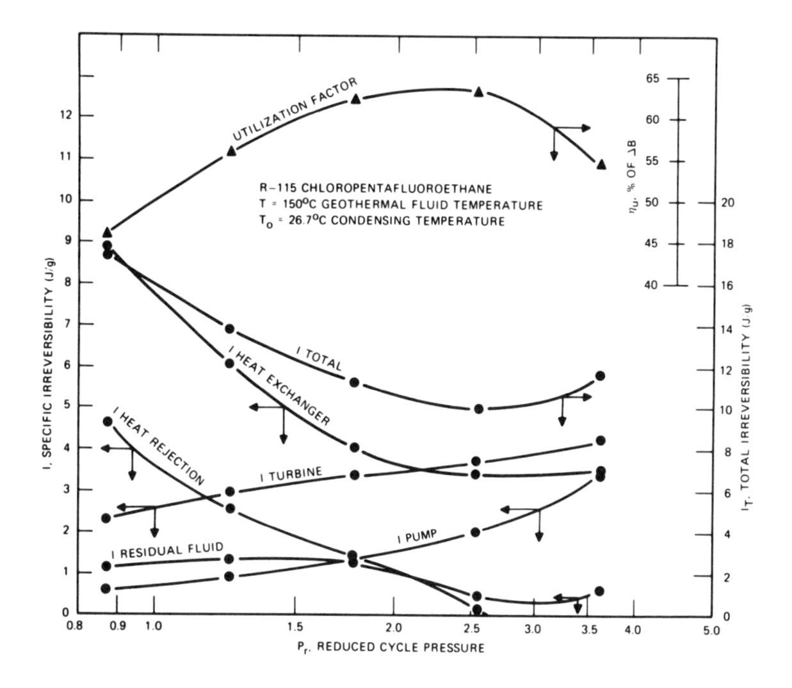

Fig. 5
Component irreversibility and η_u as a function of reduced cycle pressure for R-115 with a 150°C liquid geothermal fluid source and heat rejection at 26.7°C.

heating path of the working fluid and its performance improves only negligibly thereafter. The same is true of the residual fluid and heat rejection irreversibilities; the geothermal fluid has been cooled to a minimum value, and almost all superheat has been eliminated at the turbine exhaust. The slight increase in residual fluid irreversibility at P_r's>2.5 is due to the higher inlet working fluid temperatures at the heat exchanger associated with higher cycle operating pressures. Any further increase in pressure results in deteriorating performance since the mechanical irreversibilities of the pump and turbine are increasing functions of pressure. This deterioration is compounded by decreasing values of η_t as the turbine expansion path enters the two-phase or wet region.

4.5 Working Fluid Evaluation

In the example of the previous section, optimum performance occurs at $P_r \cong 2.5$ although utilization factors as large as 60% can be realized at $P_r \cong 1.5$ where system pressures and pumping power requirements are significantly less. In general, other working fluids exhibit a behavior similar to R-115; however, specific values of optimized η_u's and the P_r's at which they occur will be dependent largely upon the working fluid properties and the resource temperature. This point is illustrated diagrammatically in the series of optimized cycle schematics presented in Figs. 6, 7 and 8. In Fig. 6, optimum single-cycle configurations are presented for ammonia, isobutane (R-600a), and R-114 for a resource temperature of 150°C. In each case, the specified turbine inlet temperature is less than or just barely greater than the critical temperature; and optimum performance occurs at subcritical operating pressures. Supercritical configurations are possible, but inefficient, since the turbine expansion paths would begin near the top of the vapor-liquid equilibrium dome and terminate well inside the wet region where turbine performance is poor. Not surprisingly, the utilization factors are low for these examples at 46.2, 43.0, and 42.4%, respectively.

The slightly superior performance of ammonia results largely from the comparatively small amount of superheat present at the turbine exhaust. Unlike R-600a and R-114, the vapor equilibrium line for ammonia is nearly vertical at temperatures near the condensing value. This property ensures that the turbine expansion path eventually intersects the saturated vapor line where condensation can proceed. In contrast to this, the saturated vapor lines for both R-600a and R-114 exhibit a smaller positive slope near the condensing temperature. A typical turbine expansion path and the vapor equilibrium line actually diverge in this region with the result that saturation conditions cannot be attained by an adiabatic expansion from subcritical pressures. It is not difficult to show that the characteristic shape of the vapor equilibrium line is influenced primarily by the value of the reduced (or molar) ideal-gas heat capacity function (C_p^*/R). In general, compounds having complex internal molecular structures such as isobutane, RC-318, and the halogenated derivatives of ethane (R-114, R-115, etc.) will tend to increase in superheat during turbine expansions from subcritical pressures, whereas relatively simple compounds such as ammonia and the halogenated methane derivatives (R-32, R-22, etc.) will tend to lose superheat and lead to wet turbine expansions. Under identical circumstances, the latter group will yield higher values of η_u in subcritical Rankine-cycle configurations.

The effect of lower relative critical temperatures on single cycle performance is illustrated in Fig. 7 where optimum cycle configurations for R-115, R-32, and R-22 are presented for conditions similar to those of the previous example. In each example, the critical temperature is substantially below the resource temperature and consequently sufficient superheat can be added to the working fluid to allow for dry turbine expansions from supercritical pressures. Of the three, R-22 has the highest T_c and hence a lower reduced operating pressure ($P_r = 1.29$) is required to prevent the fluid from expanding into the wet region within the turbine. Heat exchanger performance is less than ideal, so $\eta_u = 55.8\%$ is the lowest of the three.

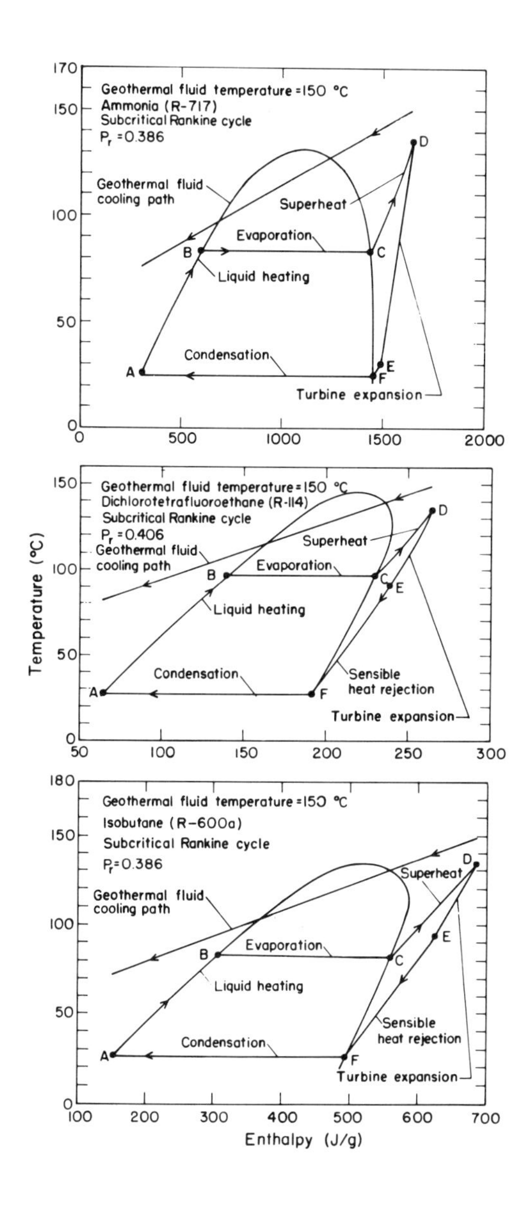

Fig. 6
Temperature-Enthalpy diagrams for subcritical Rankine cycles (100 MW(e) output) with a 150°C liquid geothermal fluid source and heat rejection at 26.7°C. Thermodynamic optimum conditions shown.

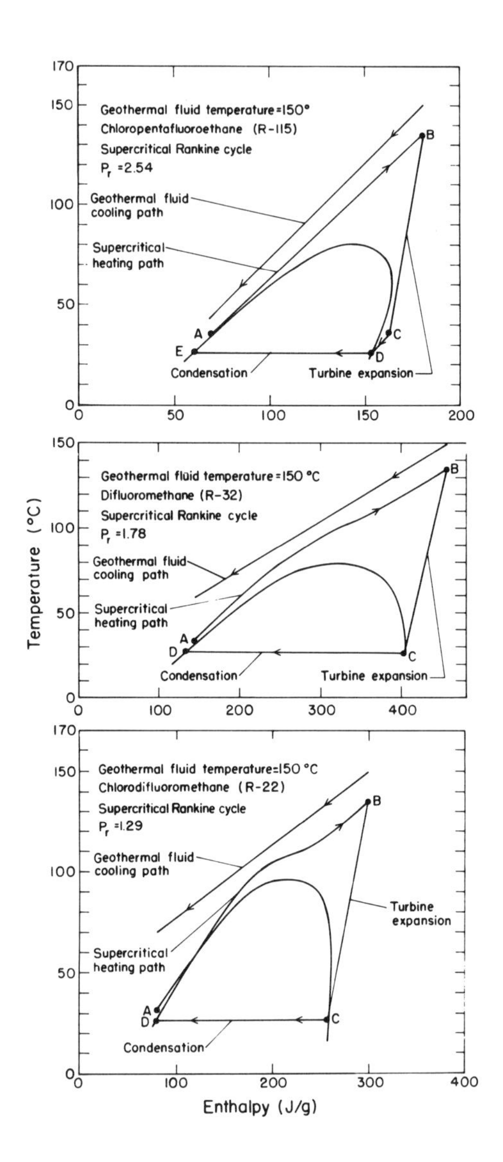

Fig. 7

Temperature-Enthalpy diagrams for supercritical Rankine cycles (100 MW(e) output) with a 150°C liquid geothermal fluid source and heat rejection at 26.7°C. Thermodynamic optimum conditions shown.

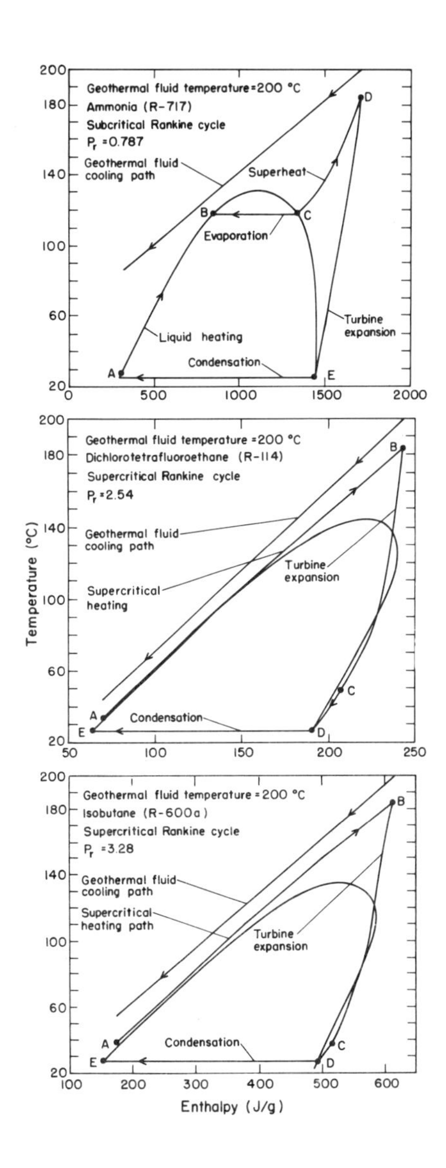

Fig. 8

Temperature-Enthalpy diagrams for Rankine cycles (100 MW(e) output) with a 200°C liquid geothermal fluid source and heat rejection at 26.7°C. Thermodynamic optimum conditions shown.

The critical temperature of R-32 is lower, and the heat exchanger performance benefits from the higher permissible reduced pressure (P_r = 1.77). Accordingly, η_u = 60.9% is an improvement relative to that of the R-22 cycle.

The critical temperature of R-115 is essentially identical to that of R-32; however, owing to the dry turbine expansion characteristics of this fluid, a much higher reduced pressure is allowed. As shown earlier in discussion of irreversibilities, heat exchanger performance is nearly ideal and a high value of η_u = 63.2% results. Fortuitously, there is only negligible superheat at the turbine exhaust (state point C). If, however, the resource temperature were higher, then state point B would move farther to the right along the P_r = 2.54 isobar with a corresponding increase in turbine exhaust superheat. To offset this, the cycle operating pressure could be increased, but doing so would introduce additional pump and turbine irreversibilities. Eventually, a temperature is reached at which these mechanical losses are dominant and the superheat cannot be removed without a corresponding decrease in utilization factor. In this manner, an optimum, or maximum, resource temperature can be identified for each working fluid. At 150°C, R-115 is at, or near, its optimum.

The effect of increasing temperature on compounds which have not reached the limit of ideal heat exchanger performance is demonstrated in Fig. 8 for ammonia, R-114 and isobutane, at a resource temperature of 200°C. Whereas these compounds were used in an inefficient subcritical mode at 150°C, the additional 50°C of resource temperature provides sufficient superheat to permit supercritical operation with isobutane and R-114. The heat exchanger performance for both fluids is excellent, and only a small amount of superheat is present at the turbine exhaust. This results in an improvement in utilization factor to nearly 70% compared to the examples of Fig. 6, which emphasizes the importance of properly matching the properties of the working fluid to the resource temperature. The dry turbine expansion characteristics of these fluids are responsible for the marked improvement. Turbine expansion paths pass briefly through the two-phase region near the top of the equilibrium dome but eventually terminate in the superheated vapor region. Turbine performance is penalized only during the part of the expansion which occurs within the two-phase region and, hence, utilization factors of 70% or more are realized. An ammonia expansion beginning at a supercritical pressure would likewise enter near the top of the equilibrium dome but would continue to expand further into the wet region resulting in unacceptably low turbine and utilization efficiencies. Consequently, a subcritical cycle is prescribed for ammonia with the optimum utilization factor being only 56%.

The effect of resource temperature on optimized performance (with respect to pressure) of several compounds is illustrated in Fig. 9 for geothermal fluid temperatures in the 100 to 300°C range. Near 100°C, all compounds operate at subcritical pressures with uniformly low utilization factors (~35%). As the temperature increases up to moderate values,

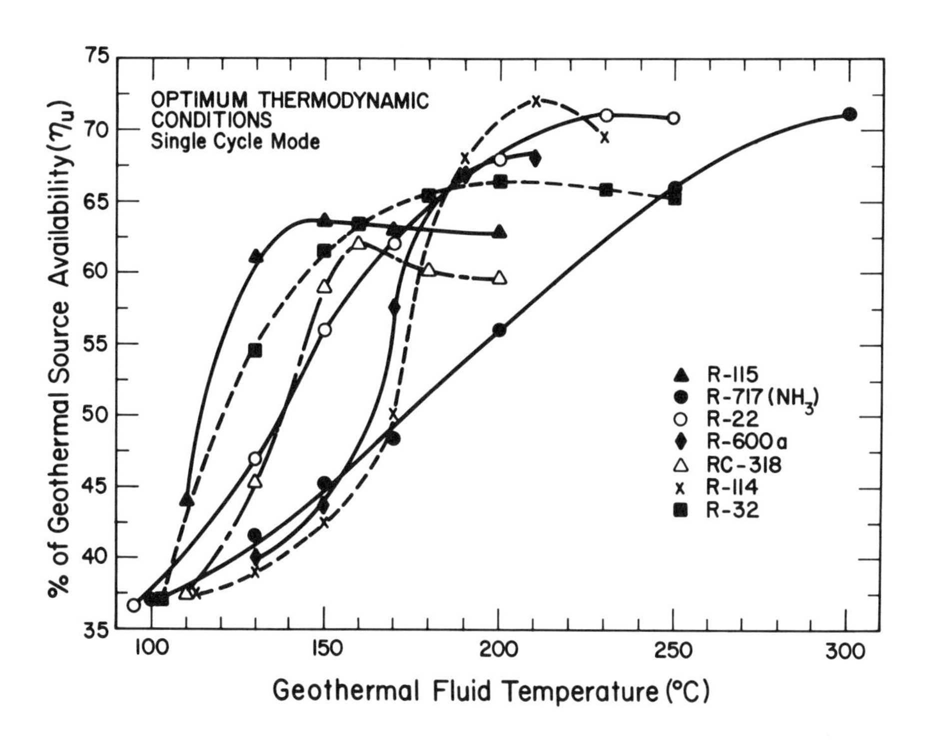

Fig. 9
Geothermal well utilization efficiency η_u as a function of geothermal fluid temperature for optimum thermodynamic operating conditions.

those compounds with lowest critical temperatures are the first to improve. Compounds having a high C_p^*/R (R-115, R-114, RC-318, R-600a) exhibit a sharp rise in η_u over a small temperature range which signifies the rapid transition to supercritical operation. On the other hand, compounds like R-22 and R-717 show a gradual improvement and attain high values of η_u over a broader range in temperature. Overall, R-22 appears to be superior at temperatures between 160 and 230°C where it attains values of η_u between 60 and 70%, respectively. Below 200°C, R-32 and R-115 ($T<150$°C) also attain high values of η_u over a relatively broad range in temperature. Ammonia is superior only above 250°C.

The exact temperature at which each compound achieves its optimum performance appears to depend primarily on the critical temperature and C_p^*/R. An empirical expression relating this characteristic temperature to these two parameters would be useful for screening candidate compounds for maximum utilization of a particular reservoir of a given temperature. Assuming that such a generalized relationship exists, the optimum temperatures shown in Fig. 9 can be used to develop the appropriate functional form. As a first approximation, one might expect that the difference between optimum resource temperature T^* and individual compound critical temperature can be expressed solely as a function of C_p^*/R:

$$T^* - T_c = f(C_p^*/R) = f\left(\frac{\gamma}{\gamma - 1}\right) . \tag{37}$$

The ratio of specific heats γ is a well-known property of most common compounds and, as such, represents a convenient way to express the quantity C_p^*/R.

In Fig. 10, the difference $T^* - T_c$ is plotted against values of C_p^*/R (or $\gamma/(\gamma - 1)$) for the seven compounds studied in Fig. 9. The data are closely approximated by the relationship

$$T^* - T_c \simeq \frac{790}{C_p^*/R} = 790\left(\frac{\gamma - 1}{\gamma}\right). \tag{38}$$

This empirical result has a simple if not intuitively plausible interpretation; at optimum performance, the molar enthalpy content of the ideal gas above the critical value is the same for all compounds (at least for those considered). This implies that compounds composed of large polyatomic molecules (γ close to unity) require very little superheat above the critical temperature to attain efficient supercritical operation while the converse is true of compounds with less complicated molecular structure such as ammonia and R-22. This explains why ammonia, whose critical temperature is below that of isobutane, performs best at a much higher temperature.

Since Eq. (38) was derived for a group of compounds with representative values of T_c and γ, it should apply as well to other compounds whose values of γ and T_c lie within the prescribed range of applicability. Table 3 lists some commonly used compounds (refrigerants) that satisfy these requirements, along with their respective values of T^* as predicted by Eq. (38). Note that a number of compounds have characteristic temperatures in the important 160 to 200°C range. In addition, most of these compounds possess relatively low values of C_p^*/R although not quite as low as R-22 and R-32. Accordingly, their characteristic performance curves would not be as broad as these two, yet neither should they exhibit the abrupt performance changes characteristic of R-114, R-115 and RC-318. Refrigerant 504, being an azeotropic mixture of R-32 and R-115, should perform like R-32 but with its optimum at a somewhat lower temperature. It would possibly make a satisfactory working fluid in the lower temperature range (150°C). Similarly, propane, R-500, R-152a, R-12, and possibly propylene should perform well in the 200°C range. At higher temperatures, R-11 and R-113 would be competitive with ammonia at least from a performance standpoint.

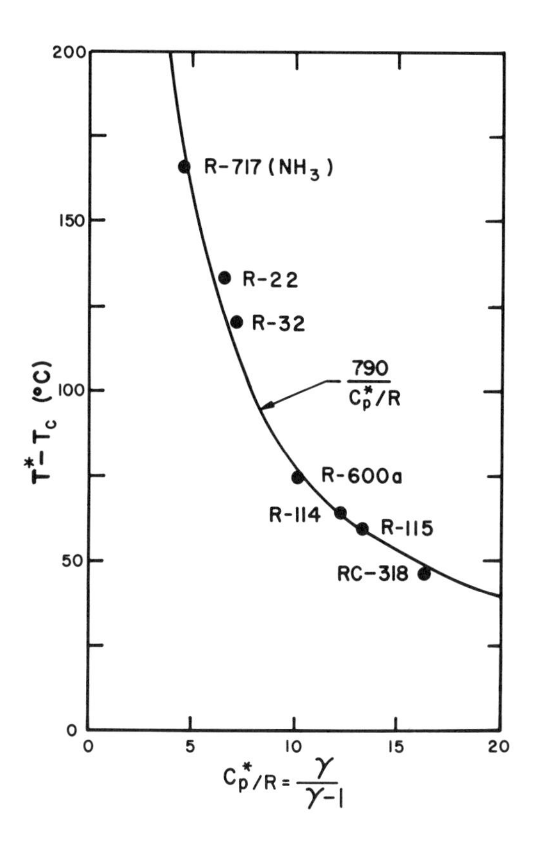

Fig. 10
Generalized correlation for the degree of superheat above the critical temperature for optimum well utilization as a function of the ideal gas state reduced heat capacity.

TABLE 3

PREDICTED VALUES OF RESOURCE TEMPERATURE FOR OPTIMUM PERFORMANCE

Compound	T_c (K)	γ	$\gamma/(\gamma - 1)$	$T^* - T_c$ (°C)	T^* (°C)
R-115	353.1	1.080	13.50	59	139
RC-318	388.5	1.066	16.15	49	164
R-13B1	340.2	1.143	8.00	99	166
R-504	339.7	1.160	7.25	109	176
R-502	363.2	1.135	8.41	94	184
R-32	351.6	1.165	7.06	112	190
R-1270 (propylene)	364.9	1.145	7.90	100	192
R-290 (propane)	370.0	1.140	8.14	97	194
R-500	378.7	1.140	8.14	97	203
R-152a	386.7	1.133	8.52	93	207
R-12	385.2	1.139	8.19	96	208
R-114	418.9	1.090	12.11	65	211
R-600a (isobutane)	408.1	1.110	10.09	78	213
R-142b	410.3	1.110	10.09	78	215
R-22	369.2	1.180	6.56	121	217
R-216	453.2	1.074	14.51	54	234
R-11	471.2	1.130	8.69	91	289
R-113	487.3	1.120	9.33	85	299
R-717 (ammonia)	405.4	1.290	4.45	178	310

Following this preliminary screening, detailed thermodynamic cycle calculations will be required for quantitative performance comparisons of at least those compounds that appear promising. For the seven compounds discussed in detail earlier, such a comparison of optimized performance is presented in Tables 4 and 5 for geothermal fluid temperatures of 150 and 200°C, respectively. The results, which include net turbine and pumping power as well as flow rates, were obtained for an assumed net generating capacity of 100 MW(e) under the conditions prescribed earlier. Also included for comparison are the well flow requirements for thermodynamically optimized dual stage flashing systems which reject heat at either 26.7 or 37.8°C. The flashing results were obtained analytically (see Appendix A) for an idealized situation that neglects parasitic losses such as additional flows required to remove noncondensable gases.

TABLE 4

COMPARISON OF OPTIMIZED PERFORMANCE OF 100 MW(e) BINARY AND TWO-STAGE FLASHING PLANTS

Geothermal fluid temperature = 150°C
Heat rejection temperature = 26.7°C

Working Fluid	Flow Rate (kg/sec)	η_u (%)	η_{cycle} (%)	Cycle Type	Net Turbine Power MW(e)	Feed Pump Power MW(e)	ΔP_r Pump Reduced Pressure Rise	P_r Maximum Cycle Pressure
R-115	1870	63.2	12.7	Supercritical	151	51	2.20	2.54
R-32	1940	60.9	13.6	Supercritical	127	27	1.49	1.78
RC-318	2040	58.8	13.1	Supercritical	116	16	1.31	1.43
R-22	2110	55.8	14.2	Supercritical	119	19	1.07	1.29
R-717 (ammonia)	2560	46.2	12.7	Subcritical	104	4	0.29	0.386
R-600a (isobutane)	2740	43.0	11.0	Subcritical	104	4	0.30	0.386
R-114	2780	42.4	12.4	Subcritical	104	4	0.34	0.406
Dual flash[a]	2182	55.0	---	---	100	0	---	---
Dual flash[b]	2598	46.0	---	---	100	0	---	---

[a]Heat rejection temperature = 26.7°C
[b]Heat rejection temperature = 37.8°C

TABLE 5

COMPARISON OF OPTIMIZED PERFORMANCE OF 100 MW(e) BINARY AND TWO-STAGE FLASHING PLANTS

Geothermal fluid temperature = 200°C
Heat rejection temperature = 26.7°C

Working Fluid	Well Flow Rate (kg/sec)	η_u (%)	η_{cycle} (%)	Cycle Type	Net Turbine Power MW(e)	Feed Pump Power MW(e)	ΔP_r Pump Reduced Pressure Rise	P_r Maximum Cycle Pressure
R-114	940	70.6	16.5	Supercritical	123	23	2.35	2.46
R-600a (isobutane)	949	68.4	17.5	Supercritical	134	34	3.14	3.28
R-22	951	68.2	17.4	Supercritical	134	34	2.41	2.63
R-32	974	66.5	17.0	Supercritical	127	27	2.24	2.54
R-115	1020	63.4	15.1	Supercritical	143	43	3.32	3.58
RC-318	1090	59.5	14.8	Supercritical	123	23	2.79	2.92
R-717 (ammonia)	1160	55.9	18.0	Subcritical	107	7	0.69	0.787
Dual flash[a]	1093	59.8	---	---	100	0	---	---
Dual flash[b]	1250	52.0	---	---	100	0	---	---

[a] Heat rejection temperature = 26.7°C

[b] Heat rejection temperature = 37.8°C

As expected, the results indicate that compounds with higher values of η_u have lower well flow requirements; and, overall, the flow rates at 150°C are substantially greater than those at 200°C (by as much as 100% for the high η_u examples). Pumping power requirements are highest (as much as 51 MW(e) at the low temperature) for the higher efficiency supercritical cycles and lowest for the subcritical cycles. As the following section shows, pumping power is primarily a function of the reduced pressure rise ΔP_r incurred across the feed pump and the thermodynamic cycle efficiency η_{cycle}.

Dual flashing systems are competitive only at the lower heat rejection temperature (26.7°C). Based upon estimates of equipment size (except evaporators) required to accommodate the large volume flows of steam at this temperature,a more realistic heat rejection temperature for flashing systems will be 37.8°C or perhaps higher. At this temperature, well flow rates are substantially higher; and, considering that these results are based upon the assumption of negligible losses incurred due to removal of noncondensibles, they are undoubtedly optimistic.

Our involved discussion of thermodynamically optimized cycles has been presented primarily for the purpose of illustrating the potential to realize efficient resource utilization. It is understood that for a specific situation, the economic optimum may occur at operating conditions that differ from those prescribed here. In particular, acceptable values of η_u will in many cases be attained at reduced pressures lower than those cited in Tables 4 and 5. This point is taken up quantitatively in the section on economic optimization where the effect of operating pressure on capital equipment and geothermal well cost is discussed.

4.6 Multiple Cycle Arrangements

The use of topping/bottoming or dual cycles (Fig. 1) might substantially increase conversion efficiency for certain geothermal fluid conditions. For example, if the wellhead source temperature is relatively high, approx. 300°C, a cycle employing a single fluid, such as isobutane, would be operating above its temperature range for optimum performance (see Table 3). Two secondary fluids operating over different pressure and temperature ranges might increase the overall utilization efficiency (η_u). Optimization of a multiple cycle arrangement is considerably more complex than a single binary-fluid cycle because it not only involves optimizing two separate cycles but also interfacing both cycles properly. Even though a general treatment would be difficult to present, the basic concepts can be extracted by considering particular examples.

A dual cycle utilizing steam and isobutane is being considered by the Los Alamos Scientific Laboratory (LASL) for a dry hot rock demonstration system with a pressurized geothermal water source at 280°C.[10] A saturated steam cycle extracts heat from the high temperature end (280 to 173°C) of the geothermal water loop, and a supercritical isobutane cycle removes heat from 173°C to a reinjection temperature of 55°C. Both the steam and isobutane cycles reject heat to the environment through induced-draft, air-cooled condensers at a condensing temperature of 49°C.

This dual cycle was optimized using a basis of 100 MW(e) output by following a scheme similar to that used for the single fluid cycle optimizations discussed in the previous section (see Fig. 11). However,

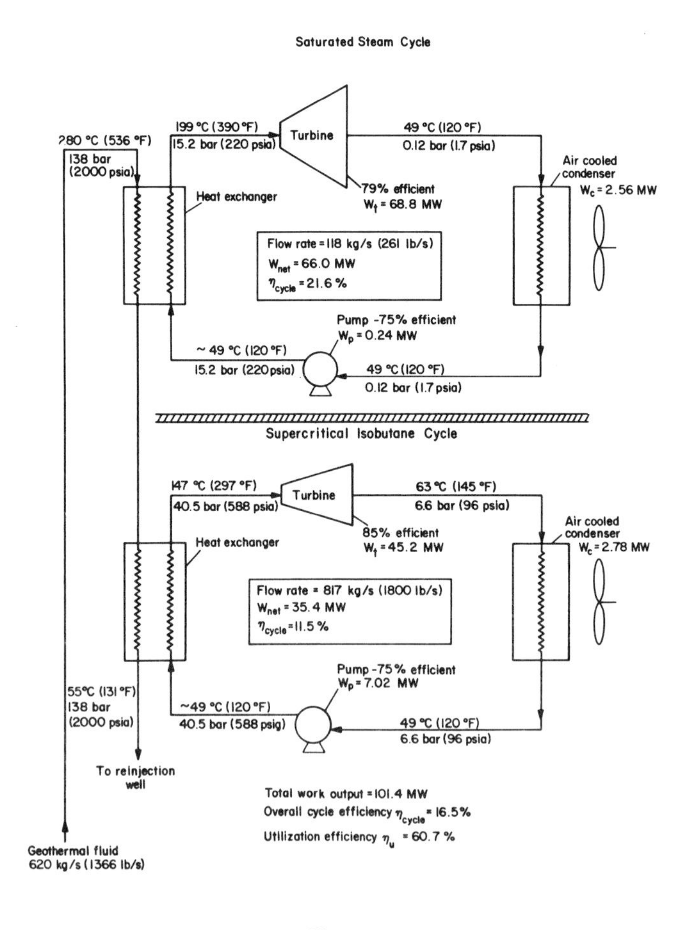

Fig. 11

Dual cycle (steam-isobutane) schematic for a 280°C pressurized water, dry hot rock geothermal source.

several important differences are brought out with the particular choice of water as the working fluid in the top end of the dual cycle. The geothermal water-steam heat exchanger is severely pinched as shown by the large temperature difference (280 vs 200°C) at the outlet. Ideally these two temperatures should be as close as is practical to improve the overall cycle efficiency. Another disadvantage is that the steam turbine operates wet beneath the liquid-vapor dome and this reduces the average turbine efficiency to 79%. Furthermore, the exhaust end size is large due to the 0.12 bar (1.7 psia) condensing pressure and consequent high specific volume at 49°C. One advantage is the relatively low pumping energy requirements due to the low liquid volumetric flow and ΔP across the pump. For example, compare the 0.24 MW at 75% efficiency pumping requirement in the steam cycle to the 7.02 MW for the isobutane cycle.

For all practical purposes, the steam cycle's performance is limited by heat exchange, in transferring heat from a sensible heat source to a latent heat sink, and by poor turbine efficiency in the saturated region. Moisture removal stages in the turbine would provide a slight improvement in efficiency and would aid in reducing turbine blade erosion. Grooved turbine blades could be used to extract condensed liquid which could then be injected at the condenser inlet. Preliminary calculations made by Landgraf, Kudrnac, and Solares [38] indicated that five stages of moisture removal would increase the net power output by 0.4 MW to 66.4 MW. Regenerative feed water heating could also be used to improve cycle efficiency. Feed water is heated prior to entering the geothermal fluid heat exchanger by direct contact with saturated steam which was extracted from the turbine at various temperatures. Landgraf and associates [38] considered the general problem of optimizing the feed water temperature and found the largest improvement in the 87 to 93°C range. For example, using one stage of feed water heating to 87°C and five stages of moisture removal results in an optimal 4% increase in power output. In practice, these gains in efficiency must be compared with the increased capital equipment and operating costs.

A topping/bottoming supercritical cycle using isobutane in both sections also seemed attractive for the 280°C proposed LASL geothermal source. The upper cycle operates from 270 to 49°C with an inlet turbine pressure of 41 bars (600 psia) and an exhaust pressure of 6.2 bars (90 psia) while the lower section operates from 200 to 49°C at 117 bars (1700 psia) turbine inlet to a 6.2 bars (90 psia) turbine exhaust pressure. The thermodynamics of the process are represented on a temperature-entropy diagram shown in Fig. 12. If the topping cycle were to operate alone its performance would be limited by excessive superheat at the turbine exhaust conditions. The addition of a bottoming cycle operating at near optimum temperature (see Table 3) recovers part of what would have otherwise been a large irreversible loss.

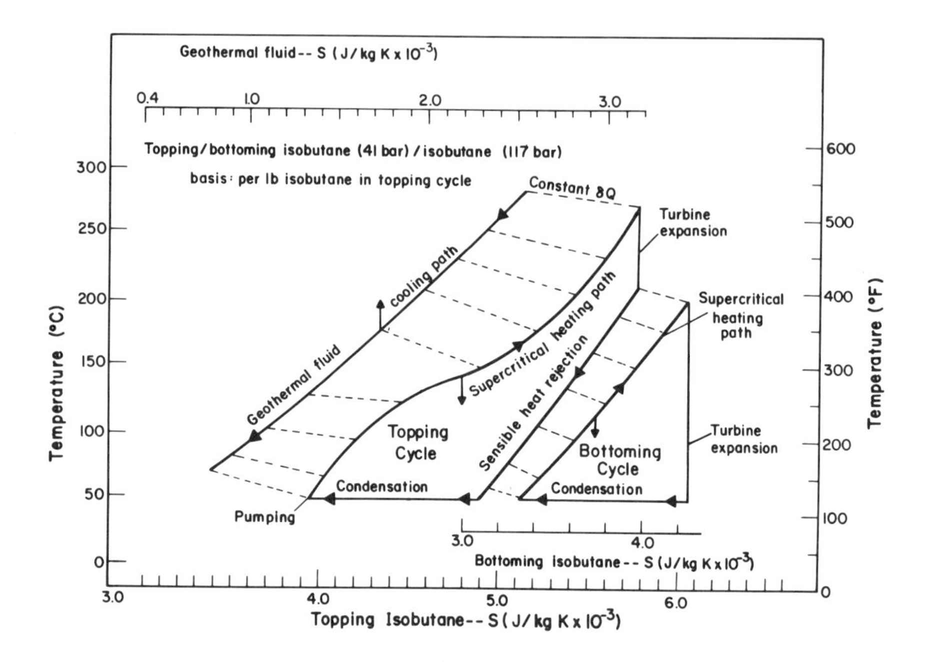

Fig. 12
Temperature-Entropy diagram for an isobutane-isobutane topping-bottoming cycle for a 280°C pressurized water dry hot rock geothermal source. Topping cycle operates at 41 bars (600 psia) and the bottoming cycle at 117 bars (1700 psia). Ideal turbine expansions shown.

Using the same geothermal fluid temperature and flow conditions that were used for a dual cycle, a 103 MW(e) net output was obtained for the topping/bottoming arrangement. Although the thermodynamic efficiencies of both cycles were almost identical, the economic picture would be much different. For instance, the two isobutane turbines would be much smaller than the steam turbine and therefore less expensive; and the condenser area requirements for the topping/bottoming arrangement would be somewhat less than the area for the dual cycle. However, the pumping requirements for the topping/bottoming cycle would be considerably greater (24.3 vs 7.3 MW). In a final analysis multiple cycle arrangements should be compared with single fluid cycles involving the same geothermal temperatures.

4.7 Brayton Gas Cycles

Brayton gas cycles differ from Rankine cycles by replacement of the condensing section with sensible heat rejection to the environment. From a heating standpoint, Brayton cycles resemble a high-pressure supercritical Rankine cycle with a practically linear temperature-enthalpy heating path in the primary heat exchanger (although the working fluid need not be at supercritical pressures). As was shown in Sec. 4.4, this is a desirable characteristic since the irreversibility associated with heat exchange is minimized when the specific heat of the working fluid is uniform. On the other hand, since waste heat is being rejected as sensible rather than latent heat, the irreversibility associated with heat rejection is high.

Brayton cycles also resemble supercritical Rankine cycles in that a large fraction of the gross turbine power is consumed in the process of compressing the gas to the cycle operating pressure. Although the pressure rise in this step may be small, the net work requirement is large due to the large value of specific volume of the gas throughout the compression. This is explained in a qualitative sense by Eq. (23) if $\bar{v}_{\ell}^{sat}$ is replaced with an appropriate average value for the gas.

The combined effects of high compressor and heat rejection irreversibilities severely restrict the energy conversion potential of Brayton cycles at these relatively low temperatures. A slight improvement in performance may be realized by operating the compression stage near the thermodynamic critical point ($P \geq P_c$, $T \approx T_c$) where the specific volume of the fluid to be compressed is comparable to that of a liquid and the pumping power is consequently lower. Only working fluids with critical temperatures below the minimum heat rejection temperature (25 to 50°C) will be suitable for this type of operation. Ethane [T_c=32°C, P_c=49.0 bars (710 psia)], ethylene [T_c=10°C, P_c=51.2 bars (742 psia)], CO_2 [T_c=31°C, P_c=73.8 bars (1070 psia)] as well as helium [T_c= −268°C, P_c=2.3 bars (33 psia)] or air [T_c= −140.6°C, P_c=37.7 bars (547 psia)] are suitable candidates, although the latter two have such low values of critical temperature that the specific volume of the compressed gas at ~50°C is large even at high pressures.

The inherently low efficiency of Brayton cycles operating at relatively low temperatures can be demonstrated by considering the same 280°C pressurized-water geothermal source with heat rejection to the environment at 49°C as was used for the dual and topping/bottoming cycle discussion. Ethylene, ethane, CO_2, and helium were selected as working fluids and the results are given in Table 6. Two cases were considered with 85 and 90% turbine and turbocompressor (pump) efficiencies assumed. Each cycle was optimized with respect to geothermal fluid utilization. Pressure ratios were selected to maximize net work produced while keeping the reinjection temperature of the geothermal fluid as low as possible. Maximum cycle pressures were chosen near the critical pressure to minimize compressor work requirements since the specific volume of the fluid is lower in this region.

In comparison to the approximately 100 MW(e) output of the dual (steam-isobutane) and topping/bottoming (isobutane-isobutane) cycles, net power outputs ranged from 0 to 38.4 MW(e). The high specific volumes involved in compressing the fluids account for the large pumping work requirements relative to water or virtually any organic liquid pumped over a similar pressure difference. In addition, the large temperature rise at the compression step associated with large ΔP's severely limits the extraction of heat from the geothermal fluid.

The importance of having high efficiencies for the turbine and compressor can be illustrated by assuming that the working fluid is an ideal gas of constant heat capacity, although the actual calculation results presented in Table 6 were based on real gas properties. The net cycle efficiency and utilization efficiency can then be expressed as an explicit function of the turbine and compressor efficiencies, η_t and η_p, the maximum geothermal fluid temperature, T_{gf}, the heat rejection temperature T_o, the expansion or compression ratio, P/P_o, and the work requirements for circulating the condenser coolant (neglecting working fluid flow losses):

$$\eta_{cycle} = \frac{W_T - W_p - W_C}{Q_{HE}} = \frac{\eta_p \eta_t \left(T_{wf}^{out}/T_o\right)\left(1 - \frac{1}{\alpha}\right) - (\alpha - 1)}{\eta_p \left(T_{wf}^{out}/T_o - 1\right) - (\alpha - 1)} - \eta_{cf} \tag{39}$$

and

$$\eta_u = \frac{\eta_{cycle}\, Q_{HE}}{C_{p_{gf}}\left[T_{gf} - T_o - T_o \ln \frac{T_{gf}}{T_o}\right]} \tag{40}$$

As η_p and η_t approach unity, η_{cycle} approaches the ideal efficiency (neglecting W_C): $1 - 1/\alpha$ which is a function only of the pressure ratio. In principle, this implies that the ideal η_{cycle} is maximized as $P/P_o \rightarrow \infty$. In practice P/P_o is limited by $P/P_o)_{max} = (T_{wf}^{out}/T)^{\gamma/\gamma - 1}$. At this pressure ratio the amount of heat extracted $\rightarrow 0$ and consequently $\eta_u \rightarrow 0$.

For a fixed maximum gas temperature T_{wf}^{out} and a fixed sink temperature T_o, the maximum work extracted from the Brayton cycle will occur at a specified optimum pressure ratio. The numerator of Eq. (39) is proportional to the net work output per unit geothermal fluid mass and can be differentiated with respect to α. The optimum pressure ratio is determined analytically to be:

$$r_{opt} = \frac{P}{P_o} = \left(\frac{T_{wf}^{out} \eta_p \eta_t}{T_o} \right)^{\frac{\gamma}{2(\gamma - 1)}} \tag{41}$$

One result is that r_{opt} is dependent on the component efficiencies η_p and η_t. For helium (γ = 1.658) with T_{wf}^{out} = 270°C and T_o = 49°C, r_{opt} varies from 1.93 to 1.26 as η_t and η_p vary from 100 to 85%. If maximum system pressures of 2500 to 3000 psia are selected this also fixes the magnitude of P_o for optimum cycle performance. For higher temperature applications, pressure ratios often deviate from r_{opt} and practical values range from 2 to 10 depending on the application and temperatures involved. Other factors such as capital cost are important. Another point worth emphasizing is that the optimum ratio for producing maximum work from a specified heat source does not correspond to a maximum cycle efficiency for a specified range of operating temperature.

The results presented in Table 6 (using real gas properties) show the significant effect that r has on the overall performance by observing changes in η_u. For example, η_u increases from 1.4 to 22.1% as r decreases from 2.86 to 1.54 (the ideal gas optimum value, Eq. (41)) for a CO_2 Brayton cycle with 85% component efficiencies. For the four fluids examined, thermodynamic optimums (maximum η_u) corresponded to ideal gas r_{opt} values. The importance of maintaining high turbine and compressor efficiencies is also illustrated by the 70 or greater percent increase in η_u as $\eta_p = \eta_t$ increased from 85% to 90%.

In general, the efficiency of geothermal heat extraction is low and thus the utilization of the geothermal fluid is limited. For CO_2, ethane, and ethylene, a saturated mixture of liquid and vapor (pressures $<P_c$) would be required at the heat rejection step to keep the heat exchanger inlet temperature at 49°C (120°F). This is due to the large temperature rise which accompanies the adiabatic compression of the fluid. If heat could be rejected at a lower temperature than 49°C, then more efficient use could be made of these fluids in a noncondensing gas cycle. Cycle efficiency can be improved by recovering some of the heat before rejection with a regenerator or recuperator; but for these geothermal fluid temperatures their performance is still too low. Unfortunately, regeneration would require a higher geothermal fluid rejection temperature resulting in poorer well utilization as well as a greater cost due to increased heat exchange surface area.

TABLE 6

THERMODYNAMICALLY OPTIMIZED NON-REGENERATIVE BRAYTON CYCLES

Geothermal Fluid Source - 280°C, 2000 psia Pressurized Water, 620 kg/sec

Working Fluid	γ	Turbine Inlet Pressure psia	Turbine Inlet Pressure MPa	Pressure Ratio r	Heat Exchanger T^{out} (°C)	Heat Exchanger T^{in} (°C)	Flow Rate (kg/sec)	Turbine MW	Gas Cooler MW	Compressor[d] MW	Net MW	η_{cycle} (%)	η_u (%)	Assumed Efficiency
Ethylene	1.24	2500	17.2	3.33	271	145	823	117.1	3.0	96.8	17.3	6.2	10.4	85% turbine & compressor efficiencies
				1.67[b]	271	77	860	56.1	3.0	21.2	31.9	6.6	19.1	
Ethane	1.19	2500	17.2	4.63	271	158	775	121.0	3.0	108.1	9.9	3.6	5.9	
				1.87[b]	271	71	761	52.6	3.0	16.7	32.9	6.4	19.7	
CO_2	1.30	2000	13.8	2.86	271	154	1836	156.1	3.0	150.7	2.4	0.9	1.4	
				1.54[b]	271	84	1726	68.3	3.0	28.4	36.9	7.9	22.1	
He[e]	1.658	2000	13.8	2.00	271	163	479	267.5	2.5	270.1	<0	---	---	
				1.26[b]	271	87	504	115.9	2.5	104.9	8.5	1.8	5.1	
Ethylene	1.24	2500	17.2	3.33	271	145	823	124.0	3.0	91.5	29.5	10.6	17.7	90% turbine & compressor efficiencies
				2.25[b]	271	111	869	94.6	3.0	53.9	37.7	9.6	22.6	
Ethane	1.19	2500	17.2	4.63	271	158	775	128.1	3.0	102.1	23.0	8.3	13.8	
				2.67[b]	271	94	752	80.3	3.0	38.9	38.4	8.9	23.0	
CO_2	1.30	2000	13.8	2.86	271	154	1836	165.3	3.0	142.3	20.0	7.2	12.0	
				1.97[b]	271	103	1763	88.6	3.0	50.1	35.5	8.9	21.2	
He[e]	1.658	2000	13.8	2.00	271	156	462	273.1	2.5	259.6	11.0	4.0	6.6	
				1.97[b]	271	112	490	182.6	2.5	156.1	24.0	5.9	14.4	

[a] Approximate values - forced draft air-cooled exchangers.

[b] $r = r_{opt} = \left(\frac{T^{out}}{T_o} \eta_p \eta_t\right)^{\gamma/2(\gamma - 1)}$ (T,K).

[c] $\eta_u = W_{net} / W_{net}^{max} = W_{net} / \Delta B \Big]^{280°C,\ 2000\ psia}_{40°C,\ 14.7\ psia}$.

[d] Inlet temperature to compressor was 49°C (120°F) in all cases.

[e] Ideal gas assumed (C_p = 1.25 BTU/lb°R).

5. CRITERIA FOR TURBINES AND PUMPS

5.1 General Turbine Design Parameters

Binary-fluid cycles may offer a significant reduction in size and cost of major plant components, especially with respect to the size and/or number of turbine exhaust ends. This is due in large part to the intrinsically higher vapor pressures and hence higher vapor phase energy densities characteristic of compounds whose critical temperatures and atmospheric boiling points are substantially lower than those of water. Low pressure condensing steam turbines operating over these modest heat drops would require numerous, large turbine exhaust ends for flashing plant sizes greater than 50 MW(e). In contrast to this, a design study by Troulakis [39] has shown that efficient and compact high-pressure ammonia turbines operating from a low-temperature heat source (comparable to our low-temperature geothermal source at ~150°C) could be built in single units at power ratings approaching 100 MW(e).

Because direct steam flashing cycles do not require a primary heat exchanger as a binary-fluid cycle would, an economic premium is placed on the proper selection of a working fluid that would result in smaller, less expensive turbines as compared to the large, low-pressure steam turbines characteristic of flashing plants. This lower turbine cost combined with improved utilization of the geothermal resource (higher η_u) tends to offset the cost of the primary heat exchanger.

5.2 Similarity Analysis of Turbine Performance

Factors affecting turbine size can be determined quantitatively by considering relationships among physical parameters that determine turbine performance, and in particular, the efficiency with which thermal energy of a gas or vapor is converted into rotating mechanical work. Dimensional analysis shows that, for geometrically similar turbines, individual turbine stage efficiency is completely determined by the Reynolds and Mach numbers of the vapor and two other characteristic nondimensional numbers involving only the stage diameter, enthalpy or heat drop, the turbine rotational speed, and the volumetric flow of vapor through the turbine blade passages.[40] Because Reynolds numbers (Re = $(\rho N D_p^2/\mu)$) are greater than 10^6 and Mach numbers (Ma = V/V_c) are less than unity for most turbines, Re and Ma number effects can be neglected. In terms of the two remaining similarity parameters, a universal functional relationship for stage efficiency can be written in the following generalized form

$$\eta_i = \eta_i(\pi_1, \pi_2) ,$$

$$\pi_1 = \frac{ND_p}{\sqrt{\Delta H_i}} , \tag{42}$$

$$\pi_2 = \frac{N^2 \dot{V}_i}{(\Delta H_i)^{3/2}}$$

where D_p is the stage pitch diameter, $\dot{V}_i$ the stage volumetric flow rate, N the turbine rotational speed, and ΔH_i the heat drop defined here as the change in specific enthalpy of the vapor as it passes through the turbine stage. The relationship implied by Eq. (42) is fundamental to the determination of optimum size (diameter) and operating characteristics (speed, volumetric flow and enthalpy drop) of turbines employing different thermodynamic working media which will result in the highest stage efficiencies. Since stage enthalpy drop and volumetric flow are closely related to the thermodynamic and physical properties of the working fluid, a turbine, designed for a particular application, such as a steam turbine, could not *a priori* be expected to perform at peak efficiency with a different fluid. More concisely, Eq. (42) implies that certain optimal values of the similarity parameters π_1 and π_2 exist at which stage efficiency will be a maximum. Optimum values can be preserved by adjusting rotational speed and stage diameter when different thermodynamic media are used.

To arrive at any realistic estimate of turbine size and operating characteristics, the thermodynamic conditions over which the turbine must operate and suitable values of the parameters π_1 and π_2 must be known. The first requirement is readily satisfied by choosing a fluid and thermodynamic cycle configuration appropriate to the characteristics of the heat source. Optimum values of π_1 and π_2 can be determined analytically by a fluid-dynamic treatment of turbomachinery performance or obtained empirically from performance data of a large number of turbines of the same type operating over a wide range of conditions, when the information exists. Actually, the former method is more expedient for single-stage performance and it is more than adequate for our purposes, as we are seeking an estimate of only the last, and largest stage diameter.

Such an analysis has been performed by Baljé[40] for various kinds of single stage turbines including the axial impulse type, commonly employed in steam power installations and the type considered here. Baljé defined his similarity parameters somewhat differently, preferring to call them specific speed, N_s, and specific diameter, D_s, both of which are related to π_1 and π_2 in the following fashion:

$$D_s = \text{specific diameter} = \frac{\pi_1}{\sqrt{\pi_2}} = \frac{D_p\left(\Delta H_i\right)^{1/4}}{\sqrt{\dot{V}_1}}$$

$$N_s = \text{specific speed} = \sqrt{\pi_2} = \frac{N\sqrt{\dot{V}_i}}{\left(\Delta H_i\right)^{3/4}} \tag{43}$$

In his analysis, Baljé determined a functional relationship of the type implied by Eq. (42) relating stage efficiency to the two similarity parameters, specific speed and specific diameter. His results show that the region of highest efficiencies (>80%) covers only a narrow range of the parameters, N_s and D_s, and hence of π_1 and π_2. Choosing from among this set of optimum values the smallest value of D_s and its corresponding N_s leads in turn to the required optimal values of the parameters π_1 and π_2. Numerically the optimum values are found to be (in consistent SI units)

$$\pi_1^{opt} = \frac{ND_p}{\sqrt{\Delta H_i}} = 0.265 \tag{44}$$

$$\pi_2^{opt} = \frac{N^2 \dot{V}_i}{\left(\Delta H_i\right)^{3/2}} = 0.0124 \ . \tag{45}$$

Equation (44) has a simple physical interpretation. The product πND_p represents the linear or peripheral speed of the turbine blades while the quantity $\sqrt{2\Delta H_i}$ is, for an impulse turbine, very nearly equal to the speed to which the gas is accelerated in the turbine nozzles, which serve the dual purpose of converting the thermal energy of the gas into kinetic energy and directing the high speed gas stream onto the stage blade row. Very simply, Eq. (44) states that for maximum conversion of the kinetic energy of the flowing gas into rotating mechanical energy, the ratio of these two speeds must be a constant, and approximately equal to one-half.

Similarly, it can be shown that using Eq. (45) is tantamount to specifying a fixed ratio of flow area through the blade passages to the total stage cross-sectional area, which is in turn equivalent to specifying the ratio of blade length to stage diameter. The latter ratio has an approximate value of one-tenth (i.e., $h^*/D_p \simeq 0.1$).

Equation (44) expresses a well-known relationship between turbine speed and diameter: for a fixed enthalpy drop, large turbines must operate at lower rotational speeds. Alternatively, for turbines of equivalent size (diameter), the rotational speed is directly proportional to the square root of the stage enthalpy drop. Moreover, the stage enthalpy drop cannot be chosen arbitrarily large, but must be limited in size to prevent sonic or choked flow conditions in the stage flow passages. This point can be illustrated by considering the expression for enthalpy drop across a stage for a hypothetical ideal fluid expanding with negligible frictional losses, a situation closely approximated in the final turbine stages where pressures are lower and ideal gas behavior is more likely to apply. Under these conditions, the enthalpy drop can be expressed analytically in terms of the stage pressure ratio, r, the gas exhaust temperature T_{ex} and the fluid properties, i.e. the heat capacity ratio $\gamma = C_p/C_v$ and the molecular weight $\mathcal{M}$.

$$\Delta H_i^{ideal} = \frac{\mathcal{R}T_{ex}}{\mathcal{M}} \left(\frac{\gamma}{\gamma - 1}\right)\left(r^{-\left(\frac{\gamma-1}{\gamma}\right)} - 1\right) \quad . \tag{46}$$

For the fluids examined, the ratio of specific heats, γ, varies between a minimum value of 1.066 (RC-318) and a maximum value of 1.3 (R-717). Within this limited range, the stage pressure ratios that would result in choked flow (sonic nozzle velocities) vary respectively between 0.592 and 0.545. These are the values of r that would give the maximum permissible enthalpy drop and require the fewest number of turbine stages. However, since sonic conditions are to be avoided, turbines must be designed to operate with stage pressure ratios greater than these minimum values. An acceptably small value of r would be 0.7 and would result in Mach numbers at the nozzle exit no greater than 0.75 to 0.8. In practice, when final design specifications are determined for a given working fluid, a somewhat lower value of r (higher stage ΔP) will probably be used, but for preliminary estimates we prefer to be conservative. For r=0.7, the terms involving γ of Eq. (46) differ over the entire range of compounds by a few percent at most, and consequently can be assumed constant and approximately equal to 0.365. The only other parameter that can significantly affect the enthalpy drop (T_{ex} and $\mathcal{R}$ are constants) is the molecular weight $\mathcal{M}$ and this can easily vary by a factor of tenfold or more,e.g. for NH_3 (R-717), $\mathcal{M}$ = 17.03 and for RC-318, $\mathcal{M}$ = 200.04.

Reemphasizing what was said earlier, for turbines of equal diameter, the rotational speed is proportional to the square root of the stage enthalpy drop and hence inversely proportional to the square root of molecular weight. It follows from Eqs. (44) and (45) that under the same conditions, the maximum permissible volumetric flow at the turbine exhaust end varies also as the inverse square root of the molecular weight. In terms of our optimized similarity parameters, the relationship among volumetric flow rate, diameter and stage enthalpy drop is given quantitatively as

$$\dot{V}_i = \sqrt{\Delta H_i}\; D_p^2 \frac{\pi_2^{opt}}{\left(\pi_1^{opt}\right)^2} = 0.177\, D_p^2 \sqrt{\Delta H_i} \;, \tag{47}$$

and for pressure ratios equal to 0.7 ($\Delta H_i \simeq 0.365\, T_{ex} \mathcal{R}/\mathcal{M}$),

$$\dot{V}_i(r = 0.7) \simeq 0.107\, D_p^2 \sqrt{\frac{\mathcal{R} T_{ex}}{\mathcal{M}}} \;. \tag{48}$$

This result along with the total exhaust flow for a particular plant determines the minimum number of exhaust ends required for a given size turbine.

5.3 Total Turbine Exhaust Flow Requirements

For any compound, the total exhaust flow is, of course, proportional to the amount of power produced and, for a given thermodynamic cycle efficiency η_{cycle}, it can be related to the amount of heat rejected from the cycle. Assuming that this heat is rejected at or near the turbine exhaust temperature and that the exhaust vapor is at saturation (there is no appreciable superheat), the following simple expression relates the total volumetric flow to the cycle efficiency, the net cycle power $\mathcal{P}$ and the thermodynamic properties of the working fluid evaluated at the turbine exhaust or heat rejection temperature, T_o.

$$\dot{V}_{total} = \left(\frac{1 - \eta_{cycle}}{\eta_{cycle}}\right) \mathcal{P} \left.\frac{v_g^{sat}}{h_{fg}}\right|_{T = T_o} \tag{49}$$

The ratio h_{fg}/v_g^{sat} is the potential volumetric energy density of the vapor and represents the capacity of the compound to carry heat at the temperature specified. This quantity must be large if turbine size and/or number of exhaust ends are to be minimal. Cycle efficiency, which is mainly a function of source and sink temperatures, is equally important in determining exhaust flow requirements. This explains why high-temperature fossil fuel plants with efficiencies approaching 40% operate with fewer turbine exhaust ends than would low-temperature flashing geothermal systems with anticipated efficiencies of 10 to 20%. Indeed, if geothermal source temperatures were significantly higher (as they are in vapor-dominated systems), the advantages of binary-fluid systems would be questionable. The opposite is true of extremely low-temperature systems.

The effect of thermodynamic properties on volumetric flow will be considered in comparing different compounds only under the conditions of identical geothermal source and heat rejection temperatures, which result in only minor variations in cycle efficiencies. To accomplish this end, it is convenient to rewrite Eq. (49) in terms of reduced thermodynamic properties:

$$\dot{V}_{total} = \left(\frac{1-\eta_{cycle}}{\eta_{cycle}}\right) P \frac{Z_c}{P_c} \left[\frac{v_g)_r^{sat}}{h_{fg})_r}\right] , \tag{50}$$

where we have introduced the concept of reduced specific volume and reduced latent heat of vaporization defined as:

$$v_g)_r \equiv \frac{v_g}{v_c} \tag{51}$$

and,

$$h_{fg})_r \equiv \frac{h_{gf} M}{RT_c} . \tag{52}$$

We will refer to the ratio $h_{fg})_r/v_g)_r^{sat}$ as the reduced volumetric energy density of the vapor. The critical compressibility factor Z_c, which appears in Eq. (50), is a property of the fluid and represents the compressibility factor $Pv\mathcal{M}/\mathcal{R}T$ evaluated at the critical point P_c, v_c, T_c. Values of Z_c for most compounds lie within the limited range 0.24 to 0.29 with the majority having values very nearly equal to 0.27. This constancy of Z_c is consistent with the corresponding-states principle which was discussed earlier. It implies a similitude of thermodynamic properties which can provide valuable insight into many problems associated with the selection of a working fluid, especially when investigating potentially attractive compounds for which little thermodynamic data are available. In particular, if similitude applies, the ratio $v_g)_r^{sat}/h_{fg})_r$ will be a function of only reduced temperature $T_r = T/T_c$. A dimensionless volumetric energy density function $E_r(T_r)$ can then be defined as:

$$v_g\big)_r^{sat}/h_{gf}\big)_r \equiv \frac{1}{E_r\left(T_r\right)}$$

and

$$\left.v_g\big)_r^{sat}/h_{fg}\big)_r\right]_{T_o} = \frac{1}{E_r\left(T_o/T_c\right)} \quad . \tag{53}$$

The exact dependence of $E_r(T_r)$ on T_r could be obtained by averaging data from many compounds. In fact, a number of generalized correlations already exist that could serve this purpose, the most common and widely used being the reduced thermodynamic properties tables by Lydersen et al.,[24] which were derived from an extensive body of thermodynamic property data. It was from these tables that we determined the form of our hypothetical energy density function. The reciprocal of this function is plotted in Fig. 13 as a function of reduced temperature. Also plotted are individual points representing the experimental (Table 2, and ref 27) values of $v_g)_r^{sat}/h_{fg})_r$ for several representative fluids evaluated at reduced temperatures corresponding to the condenser temperature T_o. With the exception of the water vapor data, the agreement between the proposed correlation and the experimental data is excellent; thus the energy density of potential working fluids can be represented as a unique function of reduced temperature. Furthermore, the generalized correlation implies an optimum reduced temperature of approximately 0.96. For a condensing temperature of 300 K (80°F), this corresponds to a critical temperature of 312 K (101°F). In practice, one would probably not employ fluids with such low critical temperatures because of the sharp reduction in energy density that would accompany slight increases in condensing temperature. In fact, the energy density approaches zero as the condensing temperature approaches the critical temperature of the fluid since the latent heat of vaporization vanishes at the critical point.

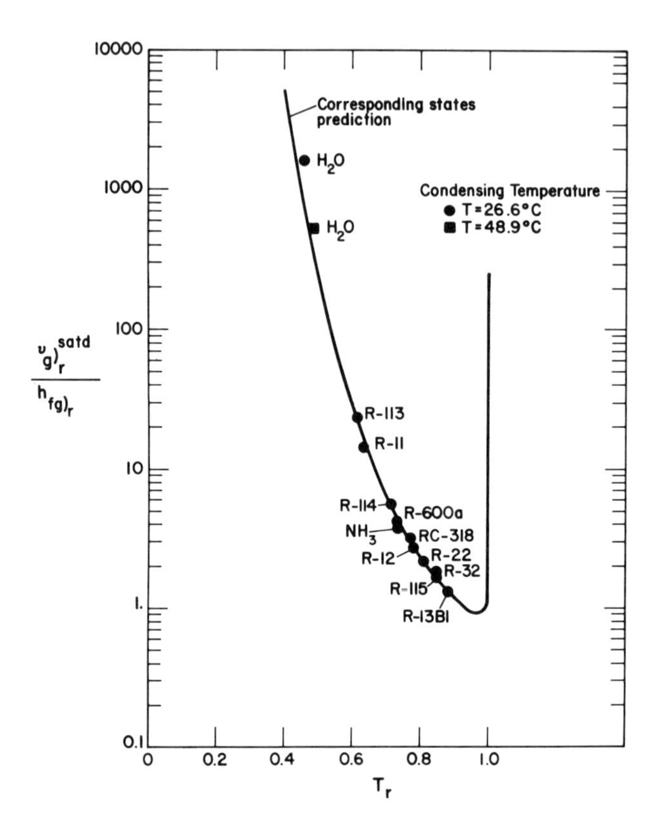

Fig. 13
Generalized correlation for the ratio of the reduced gas specific volume at saturation to the reduced latent enthalpy $(v_g)_r^{sat}/h_{fg})_r$ as a function of reduced temperature T_r.

Further examination of Fig. 13 quantitatively emphasizes the disparity in energy density between steam (T_c = 647 K) and the vapors of compounds whose critical temperatures are significantly lower. The implication is that condensing steam turbines will be much larger than turbines employing other fluids of the same power rating. In practice, steam turbine equipment is kept at an economical size by operating steam condensers at higher temperatures, typically around 322 K (120°F), and thereby accepting the penalty of reduced utilization efficiencies. Direct flashing geothermal plants, larger than demonstration size, may not be able to take advantage of low heat rejection temperatures to increase η_u. The optimum balance between utilization efficiency (well cost) and equipment cost for flashing and binary-fluid cycles will be treated later.

The reduced condensing temperature T_o/T_c, and hence the individual fluid's critical temperature are important in determining the total required exhaust flow for any given size plant. Recall also that in Eq. (50)

three other factors appear that could affect the volumetric flow: the cycle efficiency, the critical compressibility factor, and the critical pressure. Because cycle efficiencies for fluids having optimum or near optimum performance (see Section 4.5) do not vary significantly for given geothermal source and sink temperatures, then only the critical pressure is significant since values of Z_c do not differ appreciably for most of the compounds under consideration (see Table 2). Critical pressure, of course, can vary considerably as in the extreme example of water (P_c = 221.1 bars) versus RC-318 (P_c = 27.8 bars). The high critical pressure of water improves the poor performance of water vapor relative to the lower boiling compounds since it is inversely proportional to $\dot{V}_{total}$ [Eq. (50)].

The relative performance of several compounds with respect to total turbine exhaust flow is summarized in Table 7, where the quantity $v_g)_r^{sat}/h_{fg})_r]/P_c$ is tabulated for a group of compounds at their respective reduced temperatures corresponding to a condensing temperature of 26.7°C (80°F). As expected, water performs poorly in comparison to low critical temperature compounds like R-32, R-22, and R-13B1. Compounds like R-113, R-11, and R-114 have little to offer in the way of significantly lower flow rates because of their high critical temperatures and low critical pressures. Ammonia (R-717), which has an intermediate value of critical temperature, performs as well as the lower critical temperature compounds because of its relatively high critical pressure (112.8 bars). Indeed, we will show shortly that ammonia is probably the best working fluid from this standpoint, not only because total turbine exhaust flow is minimal, but also because an ammonia turbine can accommodate higher volumetric flow rates than other alternate fluid turbines of equivalent size. This high flow capacity feature obviously minimizes the number of exhaust ends required.

5.4 Sample Binary-Cycle Turbine Calculations

Comparative estimates of specific turbine requirements for a given size geothermal plant can now be made. This procedure is relatively simple because, for a given fluid and thermodynamic cycle configuration, the thermodynamic conditions (pressure, temperature, volume, and enthalpy drop) and total volumetric flow at the exhaust end are clearly established. The use of the optimized similarity parameters given by Eqs. (44) and (45) enables one to choose turbine size and volumetric flow capacity for a specified enthalpy drop. The number of exhaust ends required for an installation follows directly from individual turbine flow capacity and the known total volumetric flow.

To facilitate the selection procedure, Eqs. (44), (45), and (47) have been plotted in Figs. 14 and 15 in a way that illustrates the dependence of turbine physical characteristics ($\dot{V}_i$ and D_p) on the thermodynamic properties of the working medium (ΔH_i). Figure 14, for example, presents stage diameter as a function of enthalpy drop with rotational speed and

TABLE 7

COMPARISON OF BINARY-FLUID PLANT EXHAUST FLOW PARAMETER

$$\left.\frac{v_g)_r^{sat}}{h_{fg})_r}\right]_{T_o} \Big/ P_c, \text{ at } T_o = 26.7°C$$

Compound	T_c (K)	P_c bar	T_r (T_o = 26.7°C)	$\left.\frac{v_g)_r^{sat}}{h_{fg})_r}\right]_{T_o} \Big/ P_c$ (bar)$^{-1}$
R-32	351.6	58.3	0.853	0.031
R-13B1	340.2	39.6	0.881	0.033
R-717 (ammonia)	405.4	112.8	0.740	0.043
R-22	369.2	49.8	0.812	0.044
R-115	353.1	31.6	0.849	0.052
R-12	385.2	41.1	0.778	0.067
RC-318	388.5	27.8	0.772	0.115
R-600a (isobutane)	408.1	36.5	0.735	0.116
R-114	418.9	32.6	0.716	0.172
R-11	471.2	44.1	0.636	0.317
R-113	487.3	34.4	0.615	0.669
Water	647.3	221.2	0.463	7.233

flow capacity appearing parametrically. Figure 15 presents flow capacity also as a function of enthalpy drop, but with rotational speed and stage diameter as parameters. Both figures are equivalent and illustrate graphically that a high enthalpy drop is desirable because large volumetric flow rates can more easily be accommodated by smaller diameter turbines.

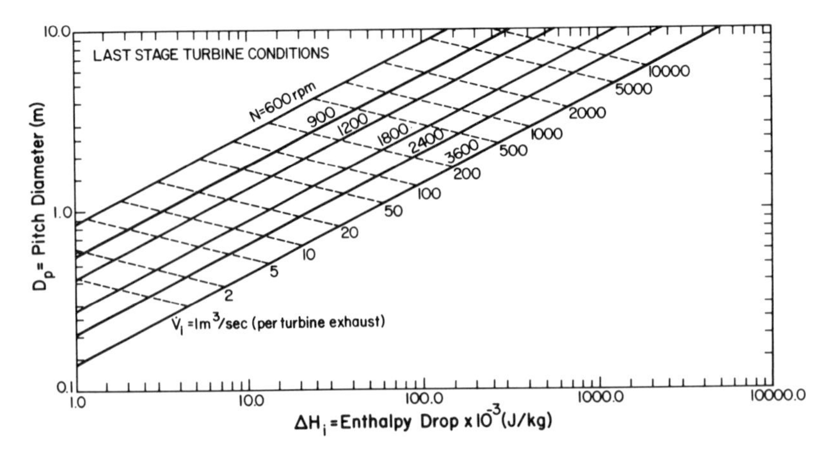

Fig. 14

Blade pitch diameter (D_p) as a function of last stage enthalpy drop (ΔH_i) for optimum turbine efficiency.

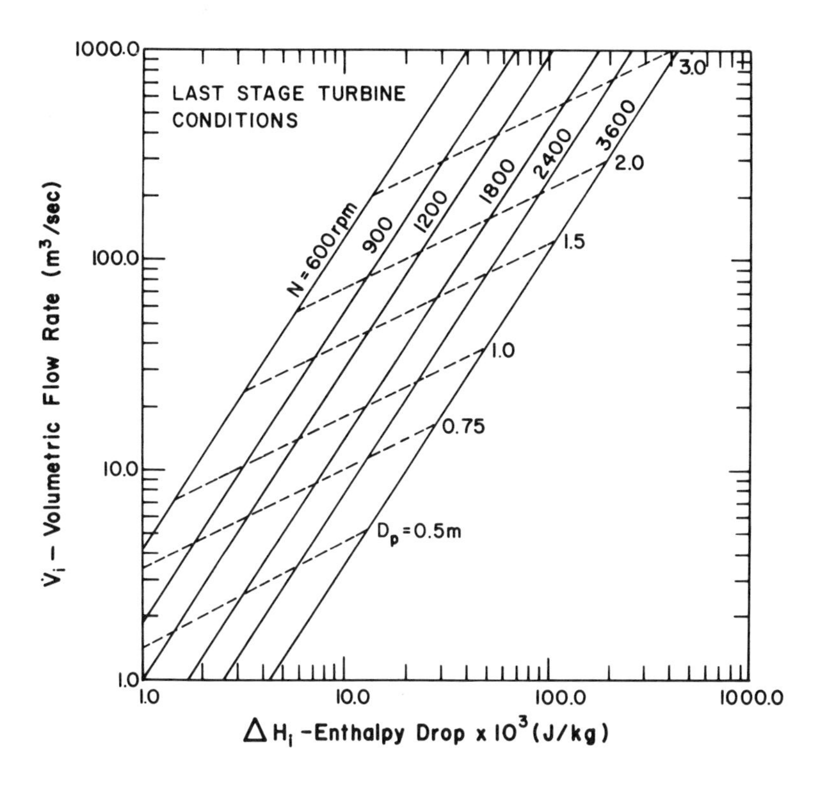

Fig. 15
Volumetric flow rate per exhaust end ($\dot{V}_i$) *as a function of last stage enthalpy drop* (ΔH_i) *for optimum turbine efficiency.*

By utilizing the fluids and thermodynamic cycle information discussed earlier in sections 4.4 and 4.5, we used Figs. 14 and 15 to calculate turbine characteristics for what might be considered as typical commerical size plants of 100 MW(e) net generating capacity. The thermodynamically optimized cycles of Figs. 6, 7 and 8 and Tables 4 and 5 were selected for study at geothermal fluid temperatures of 150 and 200°C. In each case, last stage enthalpy drops were calculated by using real fluid thermodynamic properties and Eqs. (12) and (13), an assumed dry stage efficiency of 0.85, and a stage pressure ratio of 0.7. Again, the latter value was selected to avoid sonic conditions at the nozzle exit plane.

Results of the calculations appear in Tables 8 and 9 for geothermal fluid temperatures of 150°C and 200°C respectively. Although the numbers presented can be considered representative, they are by no means unique since the total volumetric flow requirement for each fluid can be satisfied in a number of ways by various combinations of turbine size, rotational speed, and number of exhaust ends. In practice, economic considerations would presumably dictate actual turbine design.

The data of Tables 7-9 support two important claims made earlier about prospective thermodynamic working media. The first was that fluids with lower critical temperatures (R-115, R-22, R-717, and R-32) would require lower total exhaust end volume flow rates than those with higher critical temperatures (R-114, RC-318, and R-600a). Hence, fluids in the latter group tend to require numerous, large, high-capacity turbines whereas relatively few, smaller turbines are sufficient to adequately accommodate the modest flow rates required of compounds in the former group.

Secondly, it was claimed that low molecular weight compounds would allow appreciably higher enthalpy drops, which would, in turn, permit the use of compact, high-capacity turbines operating at higher rotational speeds. This explains why an ammonia turbine would have nearly twice the flow capacity as an equivalent size turbine operating with R-32 as a working fluid.

The net result of these two effects is that compounds that have a high critical temperature and/or a high molecular weight are relatively poor candidates as working fluids from the standpoint of turbine requirements. Conversely, the combination of low molecular weight and low critical temperature is desirable. As was the case for ammonia, a high critical pressure also contributes to significantly lower total exhaust flows and, hence, less severe turbine flow requirements.

In comparing the results of Table 8 with Table 9, we find, as expected, that higher source temperatures result in lower total volumetric flows due to the higher resultant thermodynamic cycle efficiencies. Whereas three ammonia turbines slightly less than 1m in diameter are required at 150°C, only two turbines of the same size will more than adequately accommodate the volume flow of 48 m^3/sec for a 100 MW(e) plant at 200°C. In fact, for an enthalpy drop of 38,175 J/kg, we see from Figs. 14 and 15 that a single 2400 rpm turbine with an exhaust diameter of only 1.3 m will easily supply the generating needs of a 100 MW(e) installation. For this and other reasons, such as its excellent condensing heat transfer characteristics, ammonia has been recognized as an excellent working fluid for a variety of prospective low-temperature energy conversion systems including fossil-fuel and nuclear power plant bottoming cycles [41,42] and ocean-thermal power plants.[43,44] Large capacity turbines operating with R-22 and R-32 are also possible although they would not be as compact as an ammonia turbine.

TABLE 8

SUMMARY OF 100 MW(e) BINARY PLANT TURBINE REQUIREMENTS

Geothermal fluid temperature = 150°C
Condensing temperature = 26.7°C

Working Fluid	Total Plant Exhaust Flow, $\dot{V}_{total}$ (m^3/sec)	Last Stage Heat Drop ΔH_i (J/kg)	Last Stage Pitch Diameter, D_p (m)	Exhaust End Flow Capacity $\dot{V}_i$ (m^3/sec)	Turbine Rotational Speed, N (rpm)	n_e Number of Exhaust Ends	Thermodynamic Cycle Configuration	Cycle Efficiency η_{cycle} (%)
R-717 (ammonia)	73	40,324	0.9	27	3600	3	Subcritical	12.7
R-32	47	10,730	0.9	16	1800	3	Supercritical	13.6
R-22	74	6,982	1.1	18	1200	4	Supercritical	14.2
R-115	94	4,634	1.2	18	900	5	Supercritical	12.7
R-600a (isobutane)	254	15,240	1.6	54	1200	5	Subcritical	11.0
RC-318	197	3,504	1.5	25	600	8	Supercritical	13.1
R-114	307	5,097	2.0	45	600	7	Subcritical	12.4

TABLE 9

SUMMARY OF 100 MW(e) BINARY PLANT TURBINE REQUIREMENTS

Geothermal fluid temperature = 200°C
Condensing temperature = 26.7°C

Working Fluid	Total Plant Exhaust Flow, $\dot{V}_{Total}$ m^3/sec	Last Stage Heat Drop, ΔH_i (J/kg)	Last Stage Pitch Diameter, D_p (m)	Exhaust End Flow Capacity, $\dot{V}_i$ (m^3/sec)	Turbine Rotational Speed, N (rpm)	n_e Number of Exhaust Ends	Thermodynamic Cycle Configuration	Cycle Efficiency η_{cycle} (%)
R-717 (ammonia)	48	38,175	0.9	27	3600	2	Subcritical	18.0
R-32	39	13,137	1.0	20	1800	2	Supercritical	17.0
R-22	58	7,040	1.4	32	900	2	Supercritical	17.4
R-115	76	4,943	1.25	20	900	4	Supercritical	15.1
R-600a (isobutane)	168	12,460	1.45	41	1200	4	Supercritical	17.5
RC-318	156	4,372	1.7	35	600	5	Supercritical	14.8
R-114	212	4,822	1.8	42	600	5	Supercritical	16.5

5.5 Proposed Method for Turbine Working Medium Evaluation

In summary, a method can be proposed whereby compounds can be quickly screened to see if they are potentially suitable working fluids from the point of view of total turbine size requirements. For a given size plant, the total exhaust flow is proportional to the group

$$\left[\frac{v_g)_r^{sat}}{h_{fg})_r}\right]_{T_o} \Big/ P_c$$

and the quantity within brackets is, for most compounds, a universal function of reduced temperature (Fig. 13). The total turbine exhaust area required for the plant is obviously proportional to the quantity $\dot{V}_{total}\, D_p^2/\dot{V}_i$, and since the ratio $D_p^2/\dot{V}_i$ was shown by Eq. (48) to be proportional to the square root of the molecular weight, it follows logically that the quantity

$$\frac{\sqrt{M}}{P_c}\left[\frac{v_g)_r^{sat}}{h_{fg})_r}\right]_{T_o}$$

is a relative measure of turbine size requirements. This quantity, defined as ξ, is a fluid property that can be considered a figure of merit; its value should preferably be small.

To evaluate ξ for any compound, one needs to know only the molecular weight, the critical pressure, the critical temperature, and the temperature at which the vapor is to condense. The ratio of the two temperatures (T_o/T_c) provides a value of reduced temperature, which is required to evaluate $v_g)_r^{sat}/h_{fg})_r$ from Fig. 13.

Values of ξ are presented in Table 10 for several commonly used compounds whose critical temperatures are above the condensing temperature of 300 K. The compounds studied in the previous section are included so that the validity of the method proposed to evaluate fluids could be checked against the results of the turbine calculations. It is not surprising that ammonia has the smallest value of ξ (0.177) due primarily to a high value of critical pressure and a small molecular weight. Refrigerants 32 and 504, with their low critical temperatures, and hence, high reduced vapor-phase energy densities, closely approach the performance of ammonia (ξ = 0.223 and 0.235, respectively). Similarly, propane and propylene have high specific energy densities and are, by far, superior to other hydrocarbons in this respect. Isobutane, although having a higher vapor phase energy density than ammonia performs

TABLE 10

FIGURE OF MERIT, ξ, FOR ESTIMATION OF RELATIVE TURBINE SIZE REQUIREMENTS

$T_o = 26.7°C$

Compound	M g/gmole	P_c (bars)	T_c (K)	$T_r(T_o = 26.7°C)$	$\left.\frac{(v_g)_r^{sat}}{(h_{fg})_r}\right\vert_{T_o}$ [(a)]	$\xi = \frac{\sqrt{m}}{P_c}\left[\frac{(v_g)_r^{sat}}{(h_{fg})_r}\right]_{T_o}$ (g/g mole)$^{1/2}$ bar^{-1}
R-717 (ammonia)	17.03	112.8	405.4	0.740	4.8	0.177
R-32	52.03	58.3	351.6	0.853	1.8	0.223
R-504	79.20	49.2	339.7	0.883	1.3	0.235
R-1270 (propylene)	42.09	47.8	364.9	0.822	1.9	0.258
R-290 (propane)	44.10	42.6	370.0	0.810	2.1	0.327
R-13B1	148.93	39.6	340.2	0.881	1.3	0.401
R-22	86.48	49.8	369.2	0.812	2.2	0.411
R-502	111.60	44.1	363.2	0.826	1.9	0.455
R-152a	66.05	46.5	386.7	0.775	2.9	0.507
R-500	99.31	45.8	378.7	0.792	2.5	0.544
R-115	154.50	31.6	353.1	0.849	1.6	0.629
R-12	120.90	41.1	385.2	0.778	2.8	0.749
R-600a (isobutane)	58.12	36.5	408.1	0.735	4.2	0.871
R-142b	100.50	42.6	410.3	0.731	4.5	1.059
R-600 (n-butane)	58.12	39.7	425.2	0.705	5.8	1.114
RC-318	200.04	27.8	388.5	0.772	3.2	1.628
R-114	170.94	32.6	418.9	0.716	5.6	2.246
R-11	137.38	44.1	471.2	0.636	14.0	3.721
R-216	220.90	28.5	453.2	0.662	10.0	5.215
R-113	187.39	34.4	487.3	0.615	23.0	9.153
Water	18.02	221.2	647.3	0.463	1600.0	30.71

[(a)] Generalized correlation used (Fig. 13)

relatively poorly because of its higher molecular weight and much smaller critical pressure. Its high critical temperature prevents it from being competitive with propane and propylene.

A few of the fluorinated hydrocarbons such as R-11, R-216 and, especially, R-113 have such high critical temperatures and molecular weights that they offer little, if any, advantage over water as potential working media for turbines. Clearly their use would be limited to small power installations of approximately 10 MW(e).

The results in Table 10 agree remarkably well with the actual turbine calculations in the previous section in the sense that relative turbine flow area requirements correlate well with the figure of merit ξ. The proposed estimation method should therefore be valuable in determining the performance of all potential working fluids without the need for numerous detailed thermodynamic cycle calculations. The effect of varying condensing temperature could also be explored easily.

5.6 Turbine Size Requirements for Geothermal Flashing Plants

Estimates of the physical characteristics of turbines were made for thermodynamically optimized two-stage flashing systems utilizing source temperatures of 150°C and 200°C. A procedure similar to that used for the binary-fluid cycle calculations was employed with one important exception. Since condensing steam turbines must typically accommodate large volumetric flows, it is necessary to increase exhaust end flow capacity, at the expense of efficiency, by allowing the ratio of blade height to pitch diameter (h^*/D_p) to be larger than the optimum value of 0.1 used in Section 5.2. For large steam turbines, blade lengths one-third as long as the last stage pitch diameter are common. The use of much smaller values of h^*/D_p would simply make the number of turbines required for moderate size flashing plants unacceptably large. For this reason, a value of one-third was used for the ratio of blade height to pitch diameter in the last stage calculations. For a given heat drop, this ratio effectively increases the exhaust end flow capacities shown in Fig. 15 by a factor of 3.33.

The total plant exhaust flows were calculated from the information presented in Appendix A which describes the optimization method for one- and two-stage flashing systems. The enthalpy drop for steam was estimated by using Eq. (46) with a stage pressure ratio of 0.7 ($\Delta H_i \approx 0.365\ \mathcal{R}T_o/\mathcal{M}$).

The results of the calculations are presented in Table 11 for two values of condensing temperature 26.7°C (80°F) and 37.8°C (100°F) to permit study of this variable's effect on total volume flow rate and the number of exhaust ends. In the worst possible case (T_{gf} = 150°C, T_o = 26.7°C), 25 turbines with a last stage pitch diameter of 2 m and blade length of 0.66 m are required to produce the necessary 100 MW(e). This number is reduced to 15 (approximately in proportion to the reduction in specific volume of saturated steam) when a condensing temperature of 37.8°C (100°F) is used. A similar result applies at 200°C; but fewer turbines are required at this temperature since the total volumetric flows are smaller.

Still, 11 large turbines are required at the higher condensing temperature with each turbine having a capacity of only 9 MW(e). The required number of turbines can be reduced further with additional increases in condensing temperature, but only at the expense of increased well cost since the small increase in condensing temperature considered here requires an additional 15% or so increase in well flow. It is desirable if possible to take advantage of lower condensing temperatures when they are available, but the enormous escalation in equipment size may prohibit this for flashing plants much larger than demonstration size. A practical lower limit to condensing temperature will probably be greater than 37.8°C (100°F).

TABLE 11

SUMMARY OF TURBINE SIZE REQUIREMENTS FOR OPTIMIZED TWO-STAGE DIRECT FLASHING PLANTS

100 MW(e) OUTPUT

Geothermal Fluid Temperature (°C)	Condensing Temperature (°C)	Total Plant Exhaust Flow (m^3/sec)	Last Stage Enthalpy Drop (J/kg)	$\dot{V}_i$ Exhaust End Flow Capacity (m^3/sec)	Last Stage Pitch Diameter (m)	n_e Number of Exhaust Ends	Rotational Speed (rpm)	Total Well Flow Rate (kg/sec)
150	26.7	12,543	50,542	500	2.0	25	1800	2182
150	37.8	7,788	52,400	532	2.0	15	1800	2598
200	26.7	9,572	50,542	500	2.0	19	1800	1093
200	37.8	5,760	52,400	532	2.0	11	1800	1250

5.7 Feed Pump Power Correlation

Rankine cycle feed pump power requirements can be estimated in a manner similar to that used to determine the approximate turbine exhaust area requirements. Assuming that heat is rejected from the thermodynamic cycle entirely as latent heat, the total plant pumping power needs are given by an expression analogous to Eq. (49)

$$P_{pump} = \frac{P\left(1 - \eta_{cycle}\right)}{\eta_{cycle}} \frac{\Delta P}{\eta_p} \left.\frac{v_\ell^{sat}}{h_{fg}}\right|_{T = T_o} , \tag{54}$$

where h_{fg}/v_ℓ^{sat} is the ratio of latent heat to specific volume of the saturated liquid and ΔP is the feed pump pressure rise. If reduced properties as defined in the preceding section are introduced, then Eq. (54) becomes:

$$P_{pump} = Z_c \frac{P}{\eta_p} \left(\frac{1 - \eta_{cycle}}{\eta_{cycle}}\right) \Delta P_r \left.\frac{v_\ell)_r^{sat}}{h_{fg})_r}\right|_{T = T_o} \tag{55}$$

where ΔP_r is the reduced pressure rise across the feed pump and is determined by the difference between the cycle operating pressure and the saturation pressure of the thermodynamic working medium at the condensing temperature, T_0.

According to the law of corresponding states, the ratio $v_\ell)_r^{sat}/h_{fg})_r$ should be a universal function of the reduced temperature, T_r (analogous to Fig. 13). This is, of course, only approximately true; but, for estimation purposes, the use of the similarity relationship between reduced thermodynamic properties is adequate. The data given in the reduced thermodynamic properties tables [24-26] can be used to establish the dependence of $v_\ell)_r^{sat}/h_{fg})_r$ on reduced temperature. In the range $0.7 \leq T_r \leq 0.9$, this ratio is adequately represented for most compounds by the simple result:

$$\frac{v_\ell\big)_r^{sat}}{h_{fg}\big)_r} = \frac{0.028}{\left(1 - T_r\right)^{2/3}} \quad . \tag{56}$$

Introducing this expression into Eq. (55) results in the following form for the generalized pumping power correlation

$$P_{pump} = 0.028\ Z_c\ \frac{P}{\eta_p}\ \frac{\left(1 - \eta_{cycle}\right)}{\eta_{cycle}}\ \frac{\Delta P_r}{\left(1 - T_r\right)^{2/3}}\Bigg|_{T_r = T_o/T_c} \tag{57}$$

The parameter which can change most and, hence, influence the pumping power most is the reduced pressure rise, ΔP_r. For the examples in Tables 4 and 5, ΔP_r assumes values between 0.294 (subcritical ammonia cycle) and 3.32 (supercritical R-115 cycle). The advantage of subcritical operation in this regard is obvious.

To a lesser degree, η_{cycle} and T_c also affect the pumping power requirement; and, in this respect, it is desirable for them to have high values. In general, as the geothermal resource temperature increases, thermodynamic cycle efficiencies improve. Furthermore, compounds that have higher critical temperatures (lower values of T_o/T_c) perform better at higher temperatures. Both of these effects reduce pumping power requirements for higher temperature operation (assuming that optimum ΔP_r's are unchanged by the changing resource temperature). The converse is true at lower resource temperatures.

The validity of Eq. (57) can be verified by comparison with the actual pumping powers obtained from Tables 4 and 5. The comparison is shown in Fig. 16, where the straight line represents the values predicted by Eq. (57). The agreement is satisfactory for all compounds except RC-318

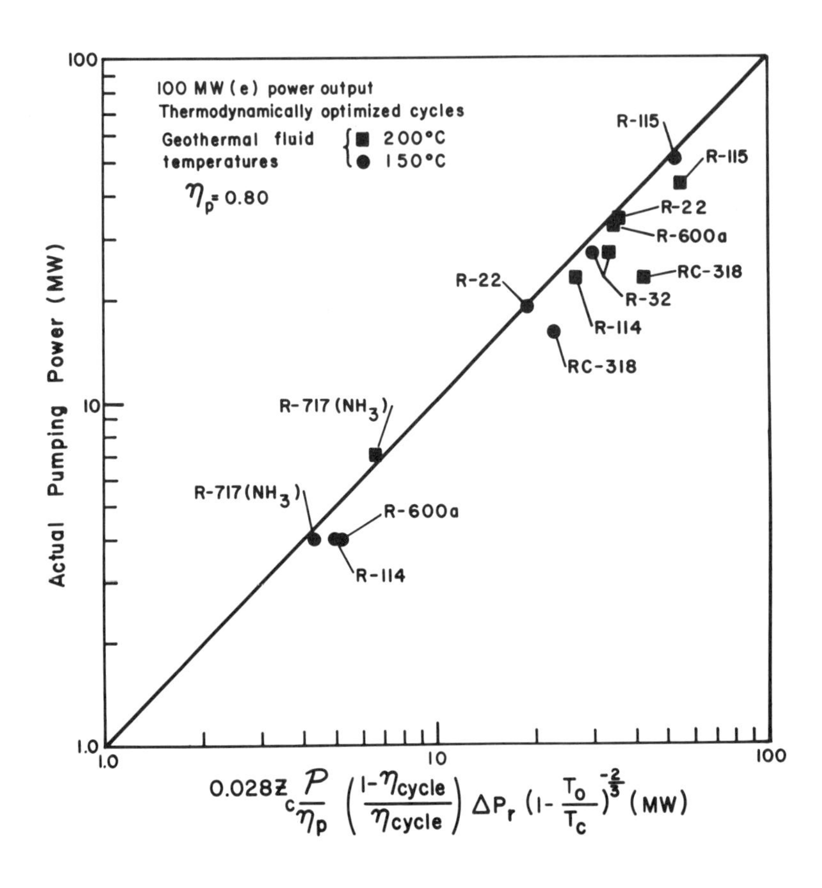

Fig. 16
Comparison between actual pumping power and theoretical pumping power based on a generalized correlation. 100 MW(e) sized plants and 150°C and 200°C geothermal fluid temperatures considered for seven working fluids.

where the discrepancy results from assuming that all heat is rejected as latent heat in the condenser. Actually, as much as 40% of the waste heat for the RC-318 cycles is rejected as sensible heat in a desuperheater. The presence of any superheat results in an overestimation of the pumping power needs. For most compounds, however, optimum thermodynamic operation occurs when the turbine expansion terminates at or near saturation conditions and then only small errors result from using Eq. (57).

6. POWER CYCLE ECONOMICS

6.1 Model Development

The total cost of a completed geothermal power plant of specified output can be estimated from the delivered cost of the major process equipment and wells. A factored estimate method was used to develop cost equations which were required in determining an economic optimum. Peters and Timmerhaus [45] and Rudd and Watson [46] show how this technique is utilized for preliminary estimates of chemical process and manufacturing plant costs. The total fixed capital investment, Φ, in the completed system can be expressed as a function of the total equipment, Φ_E, and well cost, Φ_W.

$$\Phi = \left[\Phi_E + \sum_i f_i \Phi_E\right] f_I + f_I^* \left[\Phi_W + f_W \Phi_W\right] \qquad (58)$$

The fractions f_i of Φ_E cover the costs of piping, buildings and structures, instrumentation, equipment installation, etc., and f_I is a factor that accounts for indirect expenses such as engineering fees, contingency, escalation during construction and related costs (see Tables 12 and 13). The fraction f_W of Φ_W covers the cost of piping from the wellhead to the power plant and can represent 15 to 51% of the total well cost. f_I^* includes indirect expenses associated with discovery of the geothermal field, such as land acquisition, drilling exploratory holes (one out of four successful), surface exploration, and contingencies. The details of the methodology are given in Appendices C and D.

Preliminary cost optimizations using the factored estimate method were conducted for a range of potential geothermal fluid conditions, working fluids, and cycle arrangements. Table 12 presents the fractions of the total equipment investment that were used for vapor-dominated, liquid-dominated, and artificially stimulated dry hot rock resources.

The annual costs associated with the total capital investment for plant construction depend on the interest charged on borrowed money, taxes, return on equity, depreciation, and insurance. For preliminary estimates, a fixed percentage of the total capital investment is used. A 17% figure was used in our estimates to reflect the higher interest rates that prevail today. Annual operating and maintenance costs are added to this figure in computing the cost per kWh. In cases considered, an 85% (7446 hr per year) availability or load factor was also included. To be totally realistic, the electrical transmission costs from the power plant busbar to the regional distribution network should also be added. All cost estimates are based on 1976 U.S. dollars.

TABLE 12

COST ESTIMATING FACTORS FOR GEOTHERMAL POWER PLANTS

$\Phi_E = \Phi_{HE} + \Phi_C + \Phi_T + \Phi_P$ = Equipment Total

Additional Costs (Including Labor)	Fraction of Equipment Total f_i [e] Vapor-Dominated	Liquid-Dominated	Dry Hot Rock[c]
Major equipment installation	0.10	0.10	0.10
Instrumentation	0.14	0.15	0.15
Piping (in plant)	0.10	0.08	0.08
Insulation	0.03	0.02	0.02
Foundations	0.06	0.06	0.06
Buildings & structures	0.08	0.09	0.09
Fireproofing & explosionproof equipment	0.02[a]	0.02-0.05[b]	0.02-0.05[b]
Electrical (switch yard)	0.06	0.06	0.06
Environmental controls	0.09	0.05	0.05
Total $\left(1 + \Sigma f_i\right)$	1.68	1.63-1.66	1.63-1.66

Φ_W - Production and Reinjection Wells (Drilled and Cased) = $n_{wells}\Phi^*_{well}$

Additional Costs (Including Labor)	Fraction of Equipment Total f_w [e] Vapor-Dominated	Liquid-Dominated	Dry Hot Rock[c]
Piping (well head to plant)[d]			
$n_{wells} \leq 6$	0.15	0.16	0.17
$6 < n_{wells} \leq 18$	0.23	0.24	0.25
$18 < n_{wells} \leq 36$	0.32	0.34	0.36
$36 < n_{wells} \leq 60$	0.42	0.44	0.46
$60 < n_{wells} \leq 90$	0.47	0.49	0.51
$90 < n_{wells}$	0.48	0.50	0.52

(a) Water as the working fluid.

(b) Nonflammable - 0.02, flammable - 0.05 working fluids.

(c) Assuming a pressurized-water recirculating geothermal fluid.

(d) Assuming an equilateral triangular grid with an average well spacing of 200-300 m and the power plant centrally located.

(e) Variations between liquid- and vapor-dominated and dry hot rock systems depend on factors such as insulation, pipe wall thickness, materials used, 2-phase vs 1-phase flow, gaseous effluents, and pressure losses. (See Appendices C and D).

6.2 Wells

Well cost is controlled mainly by the type of rock, diameter of the well and its depth. Studies by Altseimer [47] and Greider [48] of current cost information for oil, gas, and geothermal wells indicate a strong dependence of drilling and casing cost on depth and rock hardness. A geothermal well cost model suitable for preliminary estimating purposes was developed using the vast amount of cost information for oil and gas

TABLE 13

INDIRECT COST FACTORS FOR GEOTHERMAL POWER PLANTS[c]

Equipment related (Φ_E)	f_I	
Engineering and legal fees	0.17	
Contingency	0.13	
Overhead and escalation	0.30	
Environmental impact	0.10	
	$f_I = 1 + 0.70 = 1.70$	
Well resource discovery or exploration related $(\Phi_W)^a$	f_I^*	
Land acquisition (leasing, legal fees)	0.12 - 0.27	(0.19)[b]
Drilling exploratory holes (1 out of 4 successful)	0.09 - 0.19	(0.14)
Surface exploration (geophysical-geochemical)	0.06 - 0.14	(0.10)
Contingency	0.13	(0.13)
	$f_I^* = 1 + (0.40 - 0.73)$	= (1.56)

[a]Based on Altseimer's treatment of Greider data.[47]

[b]Represents a range of drilling costs from \$160/m to \$74/m to a depth of 1.9 km (6000 ft).

[c]See Appendix C for additional information.

wells as a basis for extrapolating the limited available data for geothermal wells. A complete description of the cost model development as well as a tabulation of well cost data are presented in Appendix C. Our approach results in drilling costs which are conservatively high and considerable savings might result in practice because of technological improvements generated by increased activity in geothermal drilling. Correlations are presented in Fig. 17 for drilling and casing cost estimates (based on 1976 dollars) in vapor-dominated, predominantly hard rock formations, liquid-dominated, softer sedimentary rock formations, and for dry hot rock stimulation wells. For a given geothermal field, an average depth should be used and multiple well costs scaled linearly. Well drilling costs are particularly difficult to estimate accurately because of the general uncertainties in the drilling operation. The costs shown in Fig. 17 should be used keeping these uncertainties in mind.

The lifetime of and production rate from a well of a given size are also difficult to estimate since the detailed character of the aquifer is usually unknown. Reasonable reservoir performance at a level greater than 50% of the original production rate has been observed for periods ranging from

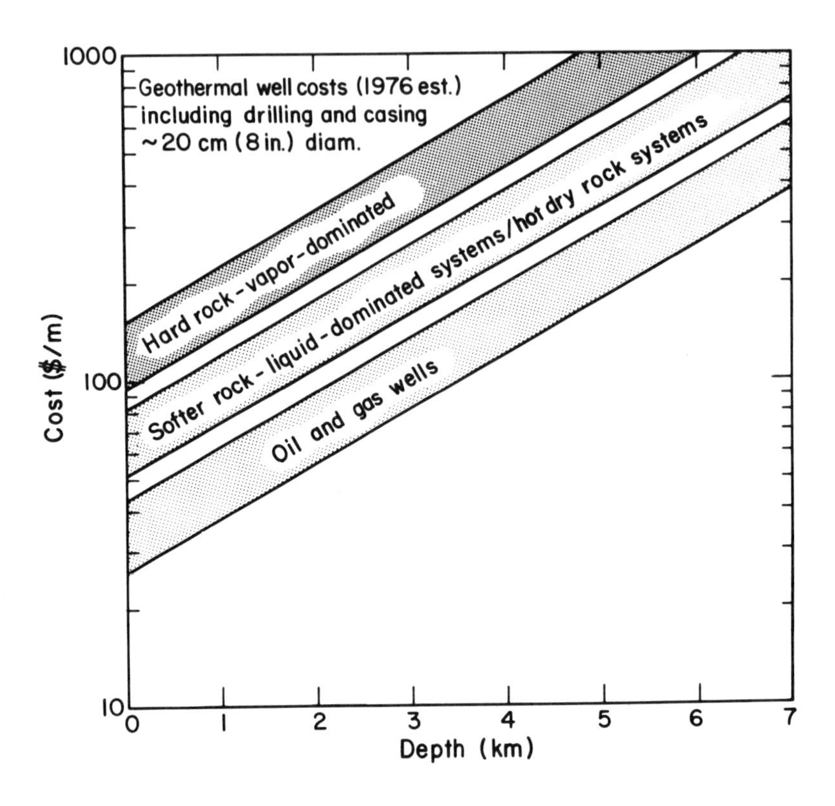

Fig. 17
Predicted well costs as a function of depth. Based on 1976 dollars.

greater than 30 years (e.g. Larderello steam fields in Italy [8]) to less than 8 years (e.g. The Geysers steam fields in California [8]). Similar results have been observed for liquid-dominated systems. Artificially stimulated, dry hot rock reservoir performance has not yet been determined from field observations, since the techniques of reservoir engineering are just being developed. Computer models have been developed at the Los Alamos Scientific Laboratory [8, p. 260] for predicting reservoir lifetime. The details of the calculational results from these modeling techniques are presented in Appendix C.

Flow rates typically range from 27 to 50 kg/sec (60 to 110 lb/sec) for artesian, liquid-dominated aquifers and from 5 to 23 kg/sec (11 to 50 lb/sec) for vapor-dominated aquifers [8] for wells of approximately 20 to 30 cm (8 to 12 in.) in diameter (see Appendix C).

Higher flows with liquid-dominated systems frequently result in premature flashing within the well bore which reduces the fluid temperature and can lead to solids precipitation and further limit subsequent well capacity. Downhole pumps not only could enhance flow but also recover the lost hydrostatic head, and thus prevent flashing that occurs in most high-temperature wells even at low flow rates.

A dry hot rock system differs considerably from the other natural geothermal systems in that the fluid flow rate from the downhole reservoir can be controlled to create a situation where conduction through the walls of the fracture limits the heat transfer. Thus by using a sufficiently large surface area system, much higher flow rates can be achieved, ~225 kg/sec (500 lb/sec), either in a self-pumping or externally-pumped mode (see Appendix C). The dry hot rock system also permits deeper drilling to obtain higher temperatures in areas of uniform temperature gradient.

Costs for both production and reinjection wells (if required) should be considered and depending on the nature of the reservoirs these might be considerably different. For example, in a dry-steam or direct-steam flashing system, part of the steam might be utilized for evaporative cooling and not returned to the aquifer. On the other hand, if a two-hole circulation system is developed for dry hot rock reservoirs, the costs for production and reinjection wells will be essentially the same. Another important factor is the uncertainty of obtaining a successful production well when exploring even in a proven geothermal field. Unsuccessful wells cost less than completed production wells because they do not need extensive surface plumbing or casing, and in some cases they serve as reinjection wells. If the geothermal area has not been selected prior to estimating the cost of a surface plant, exploration, land acquisition, and discovery costs need to be considered. These should be added to any estimate as indirect costs.

6.3 Heat Exchangers and Condensers

The cost of the heat exchangers (HE) and condensers (C) can be estimated by knowing the heat exchange surface area and the materials required. The areas A_{HE} and A_C are determined from Eq. (6) for a given size plant. Normally, a power law scaling function would be used for estimating costs, but for a 100 MW(e) plant the surface area requirements for both the primary heat exchanger and the condenser are so large (>100,000 ft^2) that there is essentially no economy of scale and multiple units of 20,000 to 35,000 ft^2 are required. Cost estimates were obtained by contacting exchanger manufacturers,and recommendations for both shell and tube and air-cooled units are presented in Fig. 18 and Appendix D.

To estimate the surface area of any exchanger, realistic values for heat transfer coefficients for both fluids, wall resistance, and fouling factors should be known. Since many different working fluids operating in both subcritical and supercritical modes are under consideration, no *uniquely* optimum design can be proposed. Furthermore, the uncertain character of the geothermal fluid itself is such that the type of heat exchanger, flow configuration, tube size, and baffling arrangements are difficult to specify *a priori*.

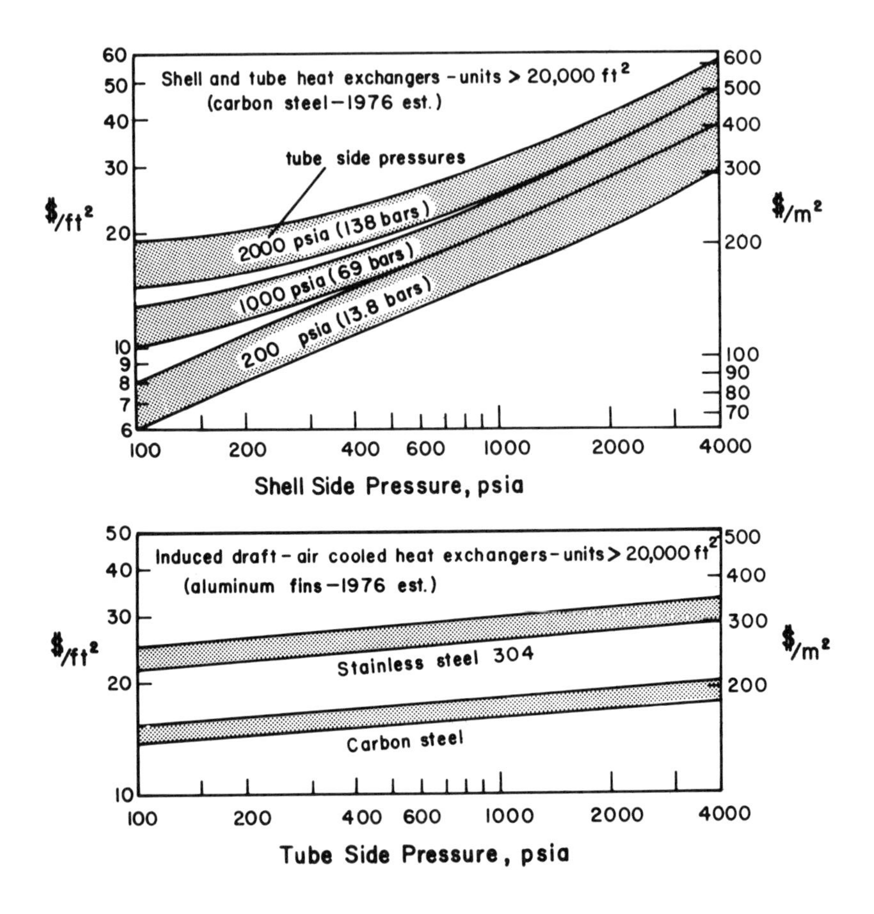

Fig. 18
Predicted heat exchanger costs as a function of shell and tube side pressure. (See Appendix D for details.) Based on 1976 dollars.

Assuming that the geothermal fluid side of the exchanger is susceptible to fouling and corrosion, geothermal fluid would be routed through the tubes to ease cleaning and reduce material costs. For liquid-dominated systems, the tube side coefficient in turbulent flow can be predicted using a Dittus-Boelter type correlation relating the Reynolds number, N_{Re}, and the Prandtl number, N_{Pr}, to the Nusselt number, N_{Nu}.[49,50]

$$N_{Nu} = \frac{hD}{k} = 0.023\ N_{Re}^{0.8} N_{Pr}^{0.33}$$

$$N_{Re} > 10{,}000 \qquad (59)$$

$$0.7 < N_{Pr} < 700 \quad .$$

For nonsupercritical flow on the shell side, similar correlations can be used for crossflow conditions on the outside of the tube bundle: (e.g. for staggered tubes with no baffle leakage [49])

$$N_{Nu} = 0.330\ N_{Re}^{0.6} N_{Pr}^{0.33} \quad . \tag{60}$$

The important difference between Eqs. (59) and (60) is that for the same N_{Re}, higher coefficients appear on the shell side because of the induced turbulence in crossflow. For fluids near the critical density ($1<\rho_r<1.5$), heat transfer coefficients are not correlated by Eqs. (59) and (60). Unfortunately, data for supercritical fluids are limited and there is no completely generalized correlation for all fluids. Studies by Bringer and Smith [51] and others [52] have indicated that a Dittus-Boelter type of equation may apply to supercritical fluids. For example, N_{Nu}^* for CO_2 flowing inside tubes has been correlated by: [52]

$$N_{Nu}^* = 0.0105\ N_{Re}^{0.875} N_{Pr}^{0.55} \quad . \tag{61}$$

Figure 19 estimates the ratio of N_{Nu}^* to N_{Nu} for crossflow over staggered banks of tubes. An adjustment to the N_{Re} scale was made to allow for the improved heat transfer in crossflow versus flow inside tubes. Although the N_{Re} corresponding to enhancement ($N_{Nu}^*/N_{Nu}>1$) is a strong function of N_{Pr}, most fluids of interest have N_{Pr}'s ranging from one to two. Thus enhancement occurs at N_{Re} greater than 10^3-10^4. For the preliminary designs considered, heat transfer coefficients were estimated conservatively low. However, more detailed designs should be based on actual data for the fluid of interest.

Because the chemical compositions and temperatures of geothermal fluids vary widely from site to site, it is extremely difficult to estimate the extent of a potential scaling problem for the primary heat exchanger unless field tests are performed. For example, the 300,000 ppm total dissolved solids (TDS) and 370°C temperatures of Salton Sea geothermal brines in the Imperial Valley present quite a different scaling problem than the geothermal brines at Cerro Prieto with 20,000 ppm TDS and 300°C temperatures.[8] Calcium carbonate ($CaCO_3$) and silica (SiO_2) scales have been observed in a number of natural geothermal systems. Dry hot rock systems introduce another factor in that the circulating water is not indigenous to the reservoir, dissolution products will build up in the solution and might also present scaling problems.

The nucleation and growth of scale deposits, particularly $Si0_2$, are complex and controlled by many factors including supersaturation, foreign ion effects (e.g. Mg^{+2}, Al^{+3}, $Fe^{+2 \text{ or } +3}$), hydrodynamic conditions (N_{Re}, rapid acceleration or deceleration of the fluid), and heat transfer rates. Nonetheless, if the scaling problem is important, then careful heat exchanger design and operation will be required. In some

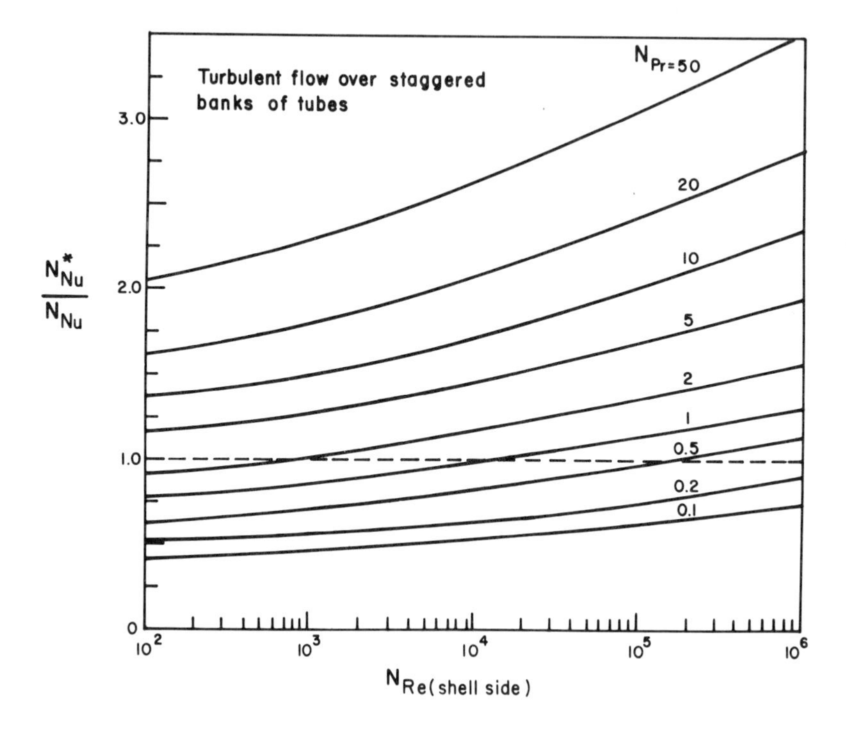

Fig. 19
The ratio of supercritical to normal fluid Nusselt numbers (N_{Nu}) as a function of Reynolds number (N_{Re}) and Prandtl number (N_{Pr}).

cases, conventional designs will be unsatisfactory and alternatives will have to be developed. Multiple stage flashing with heat exchange to an alternate working fluid, and direct contact heat exchange are possibilities.

In cases studied conventional shell and tube designs were used and correction factors were applied to the shell side coefficient for baffle-configuration, baffle-leakage, and tube bundle-bypassing effects. This typically reduced h by 25%. Fouling factors were also used and in severe cases of scaling, fouling would actually be the controlling resistance. Fouling on the geothermal fluid side was assumed to be equivalent to a coefficient of 2837 W/m^{2}K (500 BTU/hr ft^2°F). If values much lower than this are present, then other means of transferring heat should be considered, e.g. direct contact. Since the working fluid must be relatively free of contamination, an equivalent fouling coefficient of 11349 W/m^{2}K (2000 BTU/hr ft^2°F) was used. Both 2.54 cm (1 in.) and 1.90 cm (0.75 in.) tubes were considered and pressure drops limited to approximately 1 bar (14 psia). For the primary heat exchanger with supercritical flow on the shell side and pressurized geothermal fluid in the tubes, an overall coefficient U_o of 738 to 908 W/m^2K (130-160 $BTU/hrft^2 °F$) was typical. For subcritical cases, we divided the exchanger into preheat, boiling, and superheat sections and calculated coefficients applicable to each section.[50]

As the T-Q diagram for R-11 (Fig. 3) shows, the heat capacity of the working fluid changes considerably and consequently the effective temperature difference ΔT for the exchanger cannot be accurately determined using a log mean ΔT. For the cases covered in this study, effective mean ΔT_m's were calculated by graphical integration of a modified T-Q diagram.

The condenser also must be considered in some detail. In some cases, fluid at the turbine exhaust will contain a significant amount of superheat ($T>T_o$); consequently, desuperheating or sensible heat rejection is necessary before vapor condensation can take place. Overall heat transfer coefficients of 425 W/m^2K (75 $BTU/hr\ ft^2 °F$) for air-cooled and 567 W/m^2K (100 $BTU/hrft^2 °F$) for water-cooled desuperheaters would be typical. Condensing coefficients can be correlated for film-type condensation on or in horizontal tube banks by the following equation:[49,50,53]

$$h = 0.958 \left(\frac{Lk^3 \rho^2 g}{\dot{m} \mu} \right)^{1/3} = 0.725 \left(\frac{k^3 \rho^2 g\ h_{fg}}{nD\mu \Delta T} \right)^{1/4} \quad (62)$$

Because water and NH_3 have higher thermal conductivities than most of the light hydrocarbons and halogenated refrigerants, condensing coefficients vary widely (water — 8512-11348 W/m^2K (1500-2000 $BTU/hr\ ft^2 °F$), NH_3 — 5674-8512 W/m^2K (1000-1500 $BTU/hr\ ft^2 °F$), Freons and light hydrocarbons — 1135-2270 W/m^2K (200-400 $BTU/hrft^2 °F$)). Because large amounts of heat are rejected to the environment from geothermal power systems and because ΔT's are frequently limited by available ambient conditions, a large saving in condenser costs would result if the areas could be reduced by enhancing the condensing coefficient for these alternate working fluids. This is particularly true where water-cooled condensers are used because the condensing coefficient may be the controlling resistance. Fluted surfaces have enhanced steam coefficients by as much as a factor of four,[49,53] and are currently under investigation at Oak Ridge National Laboratory for their potential application to geothermal systems. In the cases treated in this study both water-cooled or finned, air-cooled condensers and desuperheaters were sized using minimal fouling on the working fluid side (U_{eq} = 11349 W/m^2K (2000 $BTU/hrft^2 °F$)) and an effective fouling U_{eq} of 5674 W/m^2K (1000 $BTU/hrft^2 °F$) on the tube side of the water-cooled systems.

In some cases of direct-steam flashing or when large amounts of desuperheating are required, spray condensers might be used. Low temperature water sources are frequently used for spray condensers with evaporative or dry cooling towers employed in the absence of such sources.[8,54] Partial removal of organic working fluid superheat can also be obtained by direct spray condensing with a stream of subcooled organic liquid.

6.4 Turbines and Pumps

Since the fluids involved (other than water) are not presently commerically used in large turbines, cost estimates were based on a model developed by the Barber-Nichols company of Denver, Colorado.[55] Turbine costs are scaled as a function of exhaust end size, D_p; blade tip speed, D_pN; number of stages, n_s; number of exhaust ends, n_e; and maximum operating pressure. Below 540°C (1000°F) temperature does not affect turbine cost significantly. Turbine and generator costs include the purchased equipment cost, and can be expressed in equation form as:

$$\Phi_T(\text{in }\$) = n_e\left(1 - 0.04\left(n_e - 1\right)\right) f_2\left[30135 n_s f_1 D_p^{2.1} + 16775 D_p^3 + 20720 D_p^2\right] + 225000\left(\frac{MW(e)}{10}\right)^{0.7} \quad (63)$$

Pitch diameters should be expressed in meters when using Eq. (63). The factor f_1 corrects for blade tip speed effects, and f_2 for pressure effects (see Fig. 20). The first term in the square bracket ($30135 n_s f_1 D_p^{2.1}$) accounts for stage costs and includes the rotor and stator charges as well as a fraction of the casing and shaft costs associated with each stage length. The second term ($16775 D_p^3$) accounts for the remainder of the casing associated with the inlet and exhaust plenums which scale as the cube of D_p. The third term ($20720 D_p^2$) covers the precision components which include labyrinth seals, radial and thrust bearings and the remainder of the shaft. The multiplier term $(1 - 0.04(n_e - 1))$ gives the economy of scale associated with multiple exhaust ends. The generator cost of the complete turbogenerator unit is represented by the term $(225{,}000\ (MW(e)/10)^{0.7})$ where MW(e) is the net electric output of the unit in megawatts. Equation (63) applies for power ratings of 1 MW to 100 MW with h^*/D_p ratios of 0.03 to 0.11, diameters up to 3 m and for $n_e \leq 4$ for tandem configurations on the same shaft.

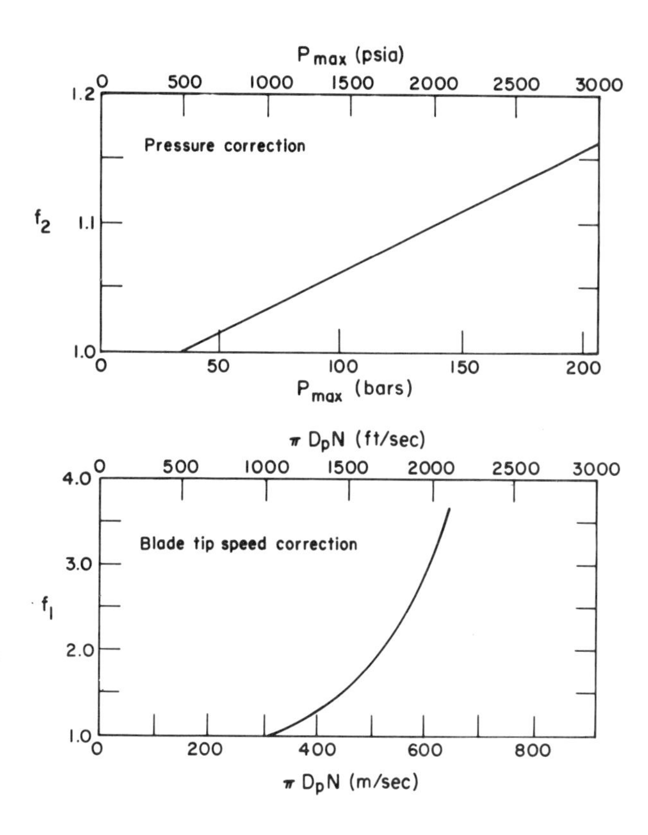

Fig. 20
Turbine cost factors for pressure and blade tip speed.

For most cases of interest employing alternate working fluids for geothermal applications, the blade tip speed will be less than 300 m/sec and f_1 will be unity. The rapid increase in f_1 for speeds greater than 300 m/sec is primarily caused by the increased cost of the higher strength alloys required. The increase in f_2 with increasing pressure is mainly due to increased casing wall thickness, and is important for many of the working fluids considered in this study.

Pump costs were based on current manufacturer estimates for large, multistage centrifugal feed pumps used in high pressure fossil-fuel fired steam electric plants (see Appendix D). Costs were scaled as a function of their power rating and are presented graphically in Fig. 21. The higher cost figure at a given power refers to high pressure operation at 138 to 207 bars (2000 to 3000 psia) while the lower cost refers to lower pressures of 34 to 69 bars (500 to 1000 psia). Because most binary-fluid cycles require feed pumping in excess of 1 MW, turbine drives rather than electric

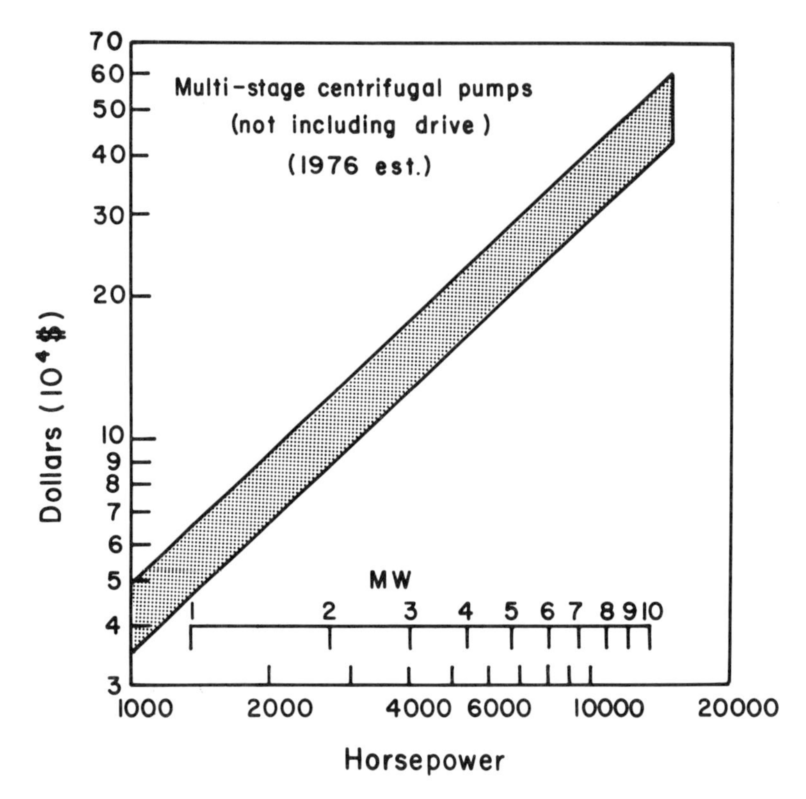

Fig. 21
Predicted pump costs not including drive mechanism. (See Appendix D for details.) Based on 1976 dollars.

motor drives would be the logical economic choice. These turbines would utilize the same fluid as was used in the main turbogenerator unit. Equation (63) without the generator term could be used for cost estimating purposes with size specifications determined by the procedures outlined in the section on turbine and pump design criteria.

6.5 Other Economic Factors

Parameters such as the power rating of the plant, the geothermal fluid inlet temperature T_{gf}^{in}, and the condensing or heat rejection temperature T_0 need to be specified before an overall cost equation can be developed. In the cases treated in this paper a 100 MW(e) power output was selected as a prototype size and T_0 ranged from 26.7°C (80°F) to 49°C (120°F). For binary-fluid cycles and direct flashing systems, the turbine inlet pressure was the primary variable since it is critical to the efficiency of the energy extraction process. Geothermal fluid temperatures ranging from 100 to 300°C were examined.

The selection of optimum ΔT's for the heat exchanger and condenser was also considered. (See Appendix B.) Large ΔT's between the geothermal and working fluid decrease the required heat exchange surface area for a given heat load, but they also decrease η_u thus requiring increased well flow for the same power output. A simplified cost model was developed to determine optimum ΔT's. An average constant C_p was assumed for both the geothermal and working fluids, and relative cost factors were applied to the major equipment components and wells. For the cases considered, a 10 to 15°C ΔT was appropriate. It should be noted that as the well cost increased relative to the major equipment component costs, the optimum ΔT decreased indicating that the amount of available energy extracted from the geothermal fluid was the critical factor. When the opposite situation existed, the optimum ΔT increased because heat exchanger area became more important.

Once a fluid (s) was selected, available thermodynamic data could be used to determine the mass and volumetric flow rates and temperature of both the geothermal and the working fluid. A procedure similar to that followed in the section on thermodynamically optimized cycles was used with the net power rating being expressed as the difference between the turbine power output and the pumping and condenser power requirements:

$$\mathcal{P} = 100\ \text{MW(e)} = \dot{W}_T - \dot{W}_P - \dot{W}_C \quad . \tag{64}$$

Using the same assumptions as before, $\mathcal{P}$ can be expressed explicitly for a binary-fluid cycle:

$$\mathcal{P} = 100\ \text{MW(e)} = \dot{m}_{wf}\left[\eta_t \left\{ H\left[P, \left(T_{gf}^{in} - 15°C\right)\right] - H\left(P_o, T_o\right)\right\}\Bigg|_S^{ideal} - \frac{\bar{v}_\ell\left(P - P_o\right)}{\eta_p} - \frac{\dot{m}_{cf}}{\dot{m}_{wf}} \eta_{cf} \Delta T_C \bar{C}_P\right] , \tag{65}$$

and the working fluid flow rate $\dot{m}_{wf}$ calculated. The mass flow rate of the geothermal fluid $\dot{m}_{gf}$ can be estimated by using the criteria established earlier, viz. a T-Q diagram with a minimum pinch point specified. In addition acceptable ΔT's would be fixed for the condenser. For each given set of independent parameters

$$\{P,\ T_{wf}^{in})_{HE},\ T_{wf}^{in})_C,\ T_{gf}^{in},\ T_o,\ P,\ P_o,\ \eta_t,\ \eta_p\}$$

a set of dependent design parameters,

$$\{\dot{m}_{gf},\ \dot{m}_{wf},\ \Delta T_{HE},\ \Delta T_C,\ U_{HE},\ U_C,\ D_p,\ n_e,\ n_s,$$

$$rpm,\ \dot{V}_{total}\ (T_o,P_o),\ v_\ell(T_o,P_o)\}$$

can be determined. These in turn can be used to estimate sizes and costs for the major pieces of equipment and wells. A cost optimum for a given set of conditions can then be determined by varying P and searching for a minimum in Φ (Eq. (58)). A few specific examples follow to illustrate the procedure.

6.6 Approach to an Economic Optimum

6.6.1 Geothermal Resources

The first case treated was a liquid-dominated 150°C geothermal resource using either direct steam flashing or a R-32 (CH_2F_2) Rankine cycle for power production. R-32 was selected as a working fluid because of its efficient resource utilization characteristics for temperatures ranging from 150 to 200°C (η_u>60%) and because of its excellent turbine figure of merit (ξ=0.223 at T_o=26.7°C(80°F)). In addition, its heat transfer characteristics should compare favorably with similar organic compounds because of its relatively high thermal conductivity. The resource was specified as a self-pumping, natural aquifer at approximately 2500 m (8202 ft) depth and limited to a maximum flow rate per well of 45 kg/sec (100 lb/sec). Well lifetimes were assumed to be greater than or equal to 20 years, the plant's estimated lifetime. Being somewhat conservative, an equal number of reinjection and production wells were specified and well costs were determined from Fig. 17 using mean values in the liquid-dominated region. The use of reinjection wells will probably be required by environmental restrictions,particularly near populated areas or where land subsidence might be a problem. Reinjection well flow rates might be substantially greater than production well flow rates for many practical systems; but for the cases treated here we assumed a lower reinjection rate on the basis that formation permeability might limit the flow somewhat. A geothermal gradient of approximately 50°C/km was also selected to be representative of a good but not exceptional resource. For example, gradients at Cerro Prieto and The Geysers are of the order of 200°C/km. The design and operating conditions employed are presented in Tables 14 and 15.

The second case was a dry hot rock resource at 250°C using a pressurized-water circulating loop and a R-717(NH_3) Rankine power cycle. Ammonia was selected as the working fluid at this temperature using the same criteria as before with R-32 at the lower temperature, namely high η_u (>60%), small turbine size (ξ=0.177 at T_o=26.7°C(80°F)), and excellent heat transfer characteristics. A vertical fracture of sufficient extent was assumed to provide a flow of 136 kg/sec (300 lb/sec) for a period of at least the 20 year lifetime of the plant. This would correspond to a 1.5 km radius crack assuming no thermal stress cracking and possibly less than a 0.78 km radius crack with a reasonable level of thermal stress cracking to enhance the heat transfer from the hot rock. The geothermal fluid flow rate of 136 kg/sec is three times greater than assumed for the natural 150°C aquifer. As mentioned earlier, higher flows are anticipated from an artificially stimulated, dry hot rock source. Furthermore, it will be interesting to compare the combined economic effects of resource temperature and well flow rate between a 250°C and 150°C geothermal system. The dry hot rock system as it is currently envisioned will involve a two-hole circulating loop;[10] and therefore the number and cost of the production and reinjection wells will be the same.

For preliminary design estimates, we assumed that neither the 150°C natural aquifer nor the 250°C dry hot rock geothermal fluids would present serious corrosion or scaling problems in heat exchangers or other process equipment. We realize that this might be an oversimplification for many geothermal applications, but in order to use currently available heat exchanger designs and costs some simplifications are necessary.* Carbon steel was employed as the material of construction for both geothermal fluid side and working fluid side applications.

*For further discussion see D.W. Shannon, "Economic Impact of Corrosion and Scaling Problems in Geothermal Energy Systems," Battelle Pacific Northwest Laboratories report BNWL-1866, UC-4 (January 1975).

TABLE 14

150°C GEOTHERMAL RESOURCE
DESIGN CONDITIONS FOR A 100 MW(e) POWER PLANT

Resource design parameters

Liquid-dominated, natural, self-pumping system

Geothermal fluid temperature T_{gf} = 150°C (300°F)

Geothermal gradient ∇T = 50°C/km

Average well depth - 2500 m (8202 ft)

Well flow rate $\dot{m}_w$ = 45 kg/sec (100 lb/sec)

Equal number of production and reinjection wells

Well lifetime $\geq$ 20 years

Power plant design parameters

R-32 binary-fluid cycle

Minimum ΔT in heat exchange or condensing steps 10°C (18°F)

Minimum heat rejection temperature T_o = 26.7°C (80°F)

Working fluid outlet temperature T_{wf}^{out} = 135°C (275°F)

Geothermal fluid reinjection temperature T_{gf}^{out} = 57.8°C (136°F)

Water-cooled condenser/desuperheater

Overall heat transfer coefficients U_o

Preheat	823 W/m^2K (145 $BTU/hr\ ft^2{}^\circ F$)
Boiling	965 W/m^2K (170 $BTU/hr\ ft^2{}^\circ F$)
Superheat	681 W/m^2K (120 $BTU/hr\ ft^2{}^\circ F$)
Supercritical	823 W/m^2K (145 $BTU/hr\ ft^2{}^\circ F$)
Desuperheat	426 W/m^2K (75 $BTU/hr\ ft^2{}^\circ F$)
Condensing	727 W/m^2K (128 $BTU/hr\ ft^2{}^\circ F$)

Turbine efficiencies 85% (dry)

Pump efficiencies 80%

Organic fluid turbine for power production and feed pump drive

Direct flashing cycles

Minimum ΔT in heat exchange or condensing steps 10°C (18°F)

Minimum heat rejection temperature T_o = 26.7°C (80°F)

Direct contact spray condenser

Low pressure steam turbines

Case 1 - exhaust to 26.7°C (80°F)

Case 2 - exhaust to 48.9°C (120°F)

6.6.2 Heat Rejection Systems

For the 150°C geothermal resource, a low temperature water supply at 10°C (50°F) was assumed for heat rejection purposes. The R-32 binary-fluid cycle would employ a shell and tube type condenser/desuperheater while the direct flashing systems might use a direct-contact spray condenser as an economical means of rejecting heat. More elaborate systems might be required and could involve cooling towers (dry or wet) which would significantly increase the equipment cost fraction of the capital investment. The 250°C case was specified to require air-cooled heat rejection corresponding to a site location in a relatively arid region of the U.S. In this instance, dry cooling towers might also be used.

6.6.3 Binary-Fluid Cycles

Basically, the effect that maximum cycle pressure had on both the total capital investment and the generating cost per kWh was examined. For the 150°C case, reduced cycle pressures from 0.87 to 3.64 for R-32 were studied while for the 250°C resource reduced pressures from 0.39 to 2.30 for NH_3 were considered. In each case, well requirements, heat exchanger and condenser/desuperheater areas, and turbine and pump sizes were determined. Equipment and well costs were estimated from the correlations presented in earlier sections. The results are graphically presented in Figs. 22-24 with a summary of design and operating conditions in Tables 14-17, and cost summaries in Table 18 corresponding to the minimum cost or economic optimum cases.

Primary heat exchange areas were determined from calculated mean ΔT's obtained by integration of the following equation:

$$A = \int \frac{dQ}{U_o \Delta T} \simeq \frac{Q_{HE}}{\bar{U}_o} \int \frac{dQ/Q_{HE}}{\Delta T} \equiv \frac{Q_{HE}}{\bar{U}_o \overline{\Delta T}_m} \quad . \tag{66}$$

The effect that changing fluid properties had on the overall heat transfer coefficient was negligible and a calculated U_o based on mean exchanger temperatures was used for all supercritical cases. For subcritical cycles, areas were calculated separately for the preheat, boiling, and superheat

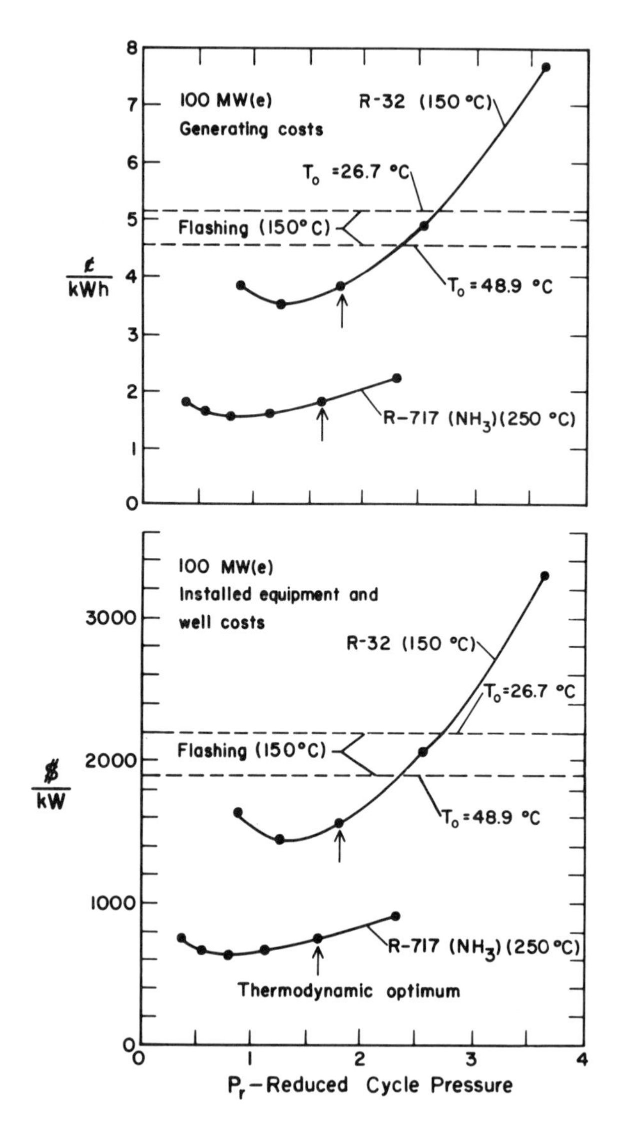

Fig. 22

Approach to economic optimum cycle conditions. Cost per kWh and cost per installed kW as a function of reduced cycle pressure P_r. NH_3 for a 250°C dry hot rock geothermal source and R-32 and direct steam flashing for a 150°C liquid-dominated geothermal source. Based on 1976 dollars.

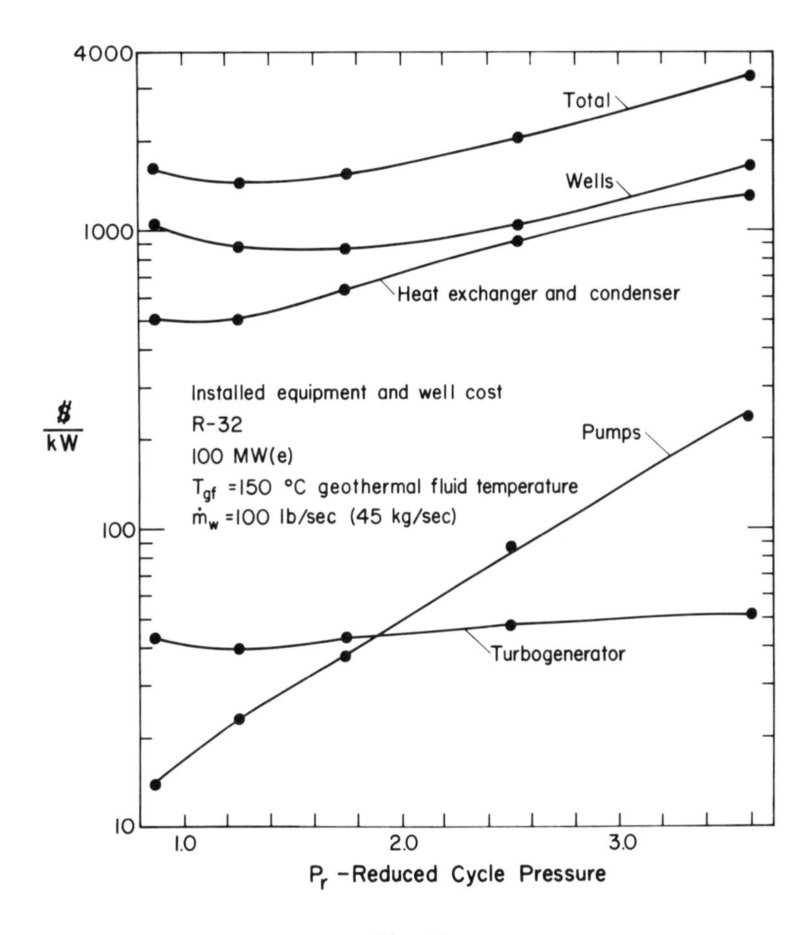

Fig. 23
Equipment and well cost breakdown for a R-32 binary-fluid cycle with a 150°C liquid-dominated geothermal source temperature and heat rejection at 26.7°C. Costs as a function of reduced cycle pressure. Based on 1976 dollars.

fractions of Q_{HE} using different U_o's and ΔT_m's (see Tables 14, 15, 16, and 17). The condenser/desuperheater was treated similarly. If sensible heat rejection was required, the exchanger area corresponding to that fraction of Q_c was calculated based on a mean ΔT and U_o for the desuperheater component. Heat exchanger and condenser costs were estimated using Fig. 18.

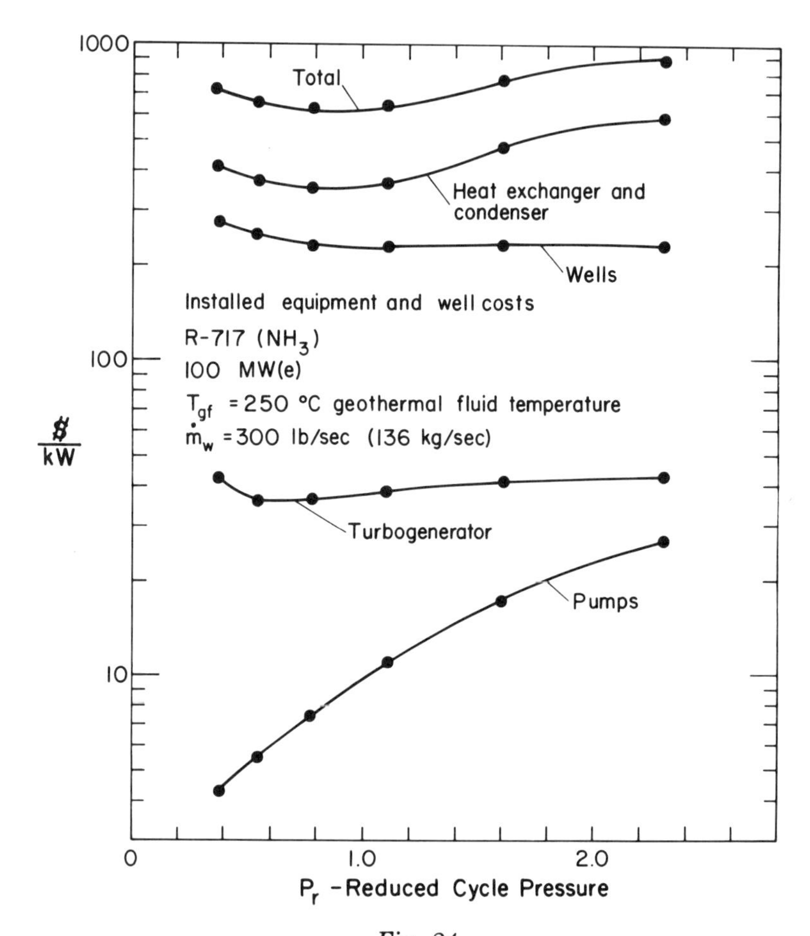

Fig. 24
Equipment and well cost breakdown for a R-717 (NH_3) binary-fluid cycle with a 250°C dry hot rock geothermal source temperature and heat rejection at 26.7°C. Costs as a function of reduced cycle pressure. Based on 1976 dollars.

Turbines were sized using Figs. 14 and 15 to ensure operation at maximum efficiency. Turbine staging was determined by a stage pressure ratio of 0.7 to prevent choking or sonic conditions. Stage efficiencies were 85% for dry conditions with a 1% per % of moisture penalty. Pump efficiencies were nominally 80% when direct organic turbine drives were used. Costs for turbogenerator and feed pump turbine drives were determined from Eq. (63) and, those for the pumps themselves, from Fig. 21.

TABLE 15

150°C GEOTHERMAL RESOURCE - OPTIMUM OPERATING CONDITIONS FOR A 100 MW(e) POWER PLANT

Item	R-32 Binary-Fluid	Direct Flashing (T_o = 26.7°C)	Direct Flashing (T_o = 48.9°C)
Geothermal fluid flow rate $\dot{m}_{gf}$	2000 kg/sec (4409 lb/sec)	2182 kg/sec (4810 lb/sec)	3267 kg/sec (7203 lb/sec)
Working fluid flow rate $\dot{m}_{wf}$	2140 kg/sec (4718 lb/sec)	Steam flow 317.7 kg/sec	417.4 kg/sec
Heat input Q_{HE}	772 MW (2.637 × 10^9 BTU/hr)		
Net power output P	100 MW (0.340 × 10^9 BTU/hr)	100 MW (0.340 × 10^9 BTU/hr)	100 MW (0.340 × 10^9 BTU/hr)
Heat rejected Q_c	672 MW (2.296 × 10^9 BTU/hr)		
Feed pump power	16 MW (21,456 hp)	~0	~0
η_u	59.2%	55.0%	36.2%[a]
η_{cycle}	12.9%	---	---
P_r	1.248 (1056 psia)	Wellhead pressure 4.6 bars (67 psia)	4.6 bars (67 psia)
Heat exchange			
Effective Mean $\overline{\Delta T}_m$			
Primary heat exchanger (supercritical)	16.1°C (29.0°F)	---	---
Desuperheater (14% of Q_c)	22.2°C (40.0°F)	---	---
Condenser (86% of Q_c)	12.2°C (22.0°F)	---	---
Primary heat exchange area	58,238 m^2 (626,873 ft^2)	---	---
Desuperheater/condenser area	75,109 m^2 (808,462 ft^2)	---	---
Turbine			
n_s - turbine stages	4	10	7
ΔH_i - last stage enthalpy drop	13,291 J/kg	50,542 J/kg	54,900 J/kg
N - rotational speed	2400 rpm	1800 rpm	1800 rpm
$\dot{V}_i$ - exhaust end capacity	12 m^3/sec (424 ft^3/sec)	500 m^3/sec (17,657 ft^3/sec)	560 m^3/sec (19,776 ft^3/sec)
V_{total} - volumetric flow	46 m^3/sec (1624 ft^3/sec)	12,543 m^3/sec (442,952 ft^3/sec)	5294 m^3/sec (186,956 ft^3/sec)
n_e - number of exhaust ends	4	25	10
D_p - blade pitch diameter	0.75 m (2.46 ft)	2.0 m (6.56 ft)	2.0 m (6.56 ft)
f_1 - tip speed factor	1	1	1
f_2 - pressure factor	1.035	1	1
h^*/D_p - blade height ratio	0.1	0.33[b]	0.33[b]

TABLE 15 (continued)

150°C GEOTHERMAL RESOURCE - OPTIMUM OPERATING CONDITIONS FOR A 100 MW(e) POWER PLANT

Item	R-32 Binary-Fluid	Direct Flashing (T_o = 26.7°C)	Direct Flashing (T_o = 48.9°C)
Feed Pump Turbine Drive			
n_s - turbine stages	4	---	---
ΔH_i - last stage enthalpy drop	13,291 J/kg	---	---
N - rotational speed	2400 rpm	---	---
$\dot{V}_i$ - exhaust end capacity	12 m^3/sec (424 ft^3/sec)	---	---
$\dot{V}_{total}$ - volumetric flow	7.3 m^3/sec (260 ft^3/sec)	---	---
n_e - number of exhaust ends	1	---	---
D_p - blade pitch diameter	0.75 m (2.46 ft)	---	---
f_1 - tip speed factor	1	---	---
f_2 - pressure factor	1.035	---	---
h^*/D_p - blade height ratio	0.1	---	---
Flashing conditions			
1st stage flash temperature	---	104.1°C (219.4°F)	113.2°C (235.7°F)
2nd stage flash temperature	---	63.2°C (145.8°F)	79.6°C (175.3°F)

[a] η_u calculated for T_o = 26.7°C

[b] h^*/D_p = 0.33 for last stage conditions

TABLE 16

250°C GEOTHERMAL RESOURCE
DESIGN CONDITIONS FOR A 100 MW(e) POWER PLANT

Resource design parameters

Dry hot rock pressurized self-pumping system

Geothermal fluid temperature T_{gf} = 250°C (482°F)

Geothermal gradient ∇T = 60°C/km

Average well depth = 3930 m (12,882 ft)

Well flow rate $\dot{m}_w$ = 136 kg/sec (300 lb/sec)

Equal number of production and reinjection wells

Well lifetime $\geq$ 20 years

Power plant design parameters

R-717 (NH_3) binary-fluid cycle

Minimum ΔT in heat exchange or condensing steps 10°C (18°F)

Air-cooled condenser/desuperheater

Minimum heat rejection temperature T_o = 26.7°C (80°F)

Overall heat transfer coefficients U_o

Preheat	1136 W/m^2K	(200 $BTU/hr\ ft^2 {}^\circ F$)
Boiling	1136 W/m^2K	(200 $BTU/hr\ ft^2 {}^\circ F$)
Superheat	681 W/m^2K	(120 $BTU/hr\ ft^2 {}^\circ F$)
Supercritical	937 W/m^2K	(165 $BTU/hr\ ft^2 {}^\circ F$)
Desuperheat	426 W/m^2K	(75 $BTU/hr\ ft^2 {}^\circ F$)
Condensing	596 W/m^2K	(105 $BTU/hr\ ft^2 {}^\circ F$)

Turbine efficiencies 85% (dry)

Pump efficiencies 80%

NH_3 turbine for power production and feed pump drive

6.6.4 Direct Flashing Systems

Cost estimates were also made for flashing systems operating at different heat rejection temperatures (26.7°C (80°F) to 48.9°C (120°F)) with the same 150°C geothermal resource used for R-32. As was discussed previously, the large specific volume of steam at turbine exhaust temperatures of approximately 26.7°C, has a dominant effect on the size and number of exhaust ends, and therefore on the economics. Consequently, we examined the effect of turbine exhaust volumetric flows in determining economic optimum operating conditions for flashing systems. In addition, the effects of the number of flashing stages and of stage pressure and temperature were studied. A preliminary study by Jensen, Ahrens, and Ho[54] indicated that multiple staging did reduce

TABLE 17

250°C GEOTHERMAL RESOURCE
OPTIMUM OPERATING CONDITIONS FOR A 100 MW(e) POWER PLANT

Item	
Geothermal fluid flow rate $\dot{m}_{gf}$	658 kg/sec (1451 lb/sec)
Working fluid flow rate $\dot{m}_{wf}$	323 kg/sec (712 lb/sec)
Heat input Q_{HE}	503 MW (1.717 × 10^9 BTU/hr)
Net power output P	100 MW (0.341 × 10^9 BTU/hr)
Heat Rejected Q_c	403 MW (1.376 × 10^9 BTU/hr)
Feed pump power	5.3 MW (7107 hp)
η_u	63.9%
η_{cycle}	19.9%
P_r	0.787 (1288 psia)
Heat Exchange	
Effective mean $\overline{\Delta T}_m$	
Primary heat exchanger	
Preheat (34% of Q_{HE}) 21.0°C (37.9°F)	
Boiler (31% of Q_{HE}) 29.8°C (53.7°F)	
Superheat (35% of Q_{HE}) 34.6°C (62.2°F)	
Desuperheater (7% of Q_c) 23.5°C (42.3°F)	
Condenser (93% of Q_c) 12.2°C (22.0°F)	
Primary heat exchange area 18,795 m^2 (202,311 ft^2)	
Desuperheater/condenser area 54,311 m^2 (584,602 ft^2)	
Turbine	
n_s - turbine stages	6
ΔH_i - last stage enthalpy drop	46,068 J/kg
N - rotational speed	3600 rpm
$\dot{V}_i$ - exhaust end capacity	34.5 m^3/sec (1218 ft^3/sec)
$\dot{V}_{total}$ - total volumetric flow	32.5 m^3/sec (1148 ft^3/sec)
n_e - number of exhaust ends	1
D_p - blade pitch diameter	0.95 m (3.12 ft)
f_1 - tip speed factor	1.0
f_2 - pressure factor	1.05
h^*/D_p - blade height ratio	0.1
Feed pump turbine drive	
n_s - turbine stages	6
ΔH_i - last stage enthalpy drop	46,068 J/kg
N - rotational speed	14,400 rpm
$\dot{V}_i$ - exhaust end capacity	1.7 m^3/sec
V_{total} - volumetric flow	1.7 m^3/sec
n_e - number of exhaust ends	1
D_p - blade pitch diameter	0.1 m (0.33 ft)
f_1 - tip speed factor	1.0
f_2 - pressure factor	1.05
h^*/D_p - blade height ratio	0.1

costs, mainly because of improved resource utilization (η_u) and of the importance of well cost to the overall capital investment. For the cases treated in this study two flashing stages were used with turbine inlet pressures and temperatures ranging from 2.76 bars (40 psia)/131°C (267°F) to 0.69 bars (10 psia)/89°C (193°F) with a saturated wellhead pressure of 4.82 bars (70 psia) at 150°C (302°F). Approximate costs of required pumps were calculated using Fig. 21 and turbine costs were estimated by taking 1.25 times the amount given by Eq. (63) to account for the large h^*/D_p ratio in the last stages. Nominal costs were also included for flash tanks and cyclone separator systems. In addition, no account was made for non-condensable gases that might be contained in the geothermal fluid. These will tend to increase the turbine back pressure and require steam ejectors to maintain acceptable exhaust pressure levels. Design and operating conditions are summarized in Tables 14 and 15 and costs are summarized in Table 18. Overall generating costs for two flashing systems are compared with those for the R-32 cycle in Fig. 22. Optimum flashing conditions were very close to those corresponding to the thermodynamic optimum (maximum η_u) conditions derived in Appendix A. The cases presented in Tables 15 and 18 and Fig. 22 are for conditions which resulted in the minimum cost for T_o = 26.7°C and 48.9°C. A flashing case corresponding to the 250°C resource was not considered in the present analysis because of potential precipitation problems associated with flashing the higher temperature resource, although a direct flashing system at 250°C would cost less than one at 150°C assuming no major geochemical problems.

6.7 Cost Comparison of Binary-Fluid and Direct Flashing Systems

The 150°C resource economic optimum occurred at a supercritical reduced cycle pressure of 1.24 with R-32 as the working fluid (Fig. 22). This is very close to the thermodynamic optimum (η_u = 60.9%) which occurred at P_r = 1.78. It is reasonable to assume that operating pressures less than those corresponding to a maximum η_u would lower the total cost because of lower heat exchanger costs associated with lower pressure operation (see Fig. 18). This is contingent on η_u not decreasing significantly from its maximum. At the economic optimum η_u has only decreased to 59.2%. Figure 23 illustrates this concept by breaking down the total cost into its components. Heat exchanger cost decreases from \$640/kW to \$505/kW as P_r decreases from 1.78 at the thermodynamic optimum to 1.24 while the well cost is essentially constant at \$880/kW over this range. At large P_r's, when η_u decreases markedly, well and pump costs begin to dominate the economics. In contrast, turbogenerator costs remained relatively constant at \$40 to \$50/kW. The turbine component cost actually increased from \$12/kW at a P_r of 1.78 to \$20/kW at a P_r of 3.64, but the 100 MW(e) generator at \$31/kW was still the major fraction of the turbogenerator unit cost.

There are similar effects when NH_3 is used as the working fluid with the 250°C resource, except that the economic optimum corresponds to a subcritical cycle at a P_r of 0.79 again near the thermodynamic optimum which occurs at P_r of 1.61. As seen on Fig. 22, the cost curve is more uniform and below the R-32 curve indicating influence of higher well flow rate and higher resource temperature. The R-32 cycle at 150°C represents a total capital investment of $1463/kW or 3.48¢/kWh versus $625/kW or 1.56¢/kWh for the NH_3 cycle at 250°C. The improvement in cycle efficiency from 12.9% for R-32 to 19.9% for NH_3 and the larger mean ΔT for NH_3 (Tables 15 and 17) reduce the required primary heat exchange area from 58,238 m^2 (626,873 ft^2) to 18,795 m^2 (202,311 ft^2) and the condenser/desuperheater area from 75,109 m^2 (808,462 ft^2) to 54,311 m^2 (584,602 ft^2). This reduction more than compensates for the increased cost per unit area ($17.5/$ft^2$ for R-32 versus $19/$ft^2$ for the NH_3 primary heat exchanger and $9/$ft^2$ for the water-cooled R-32 condenser versus $15/$ft^2$ for the air-cooled NH_3 condenser). (See Tables 15, 17, and 18.) Well cost effects are not as important as in the R-32 case. As P_r increases from 0.78 to 2.3, there is virtually no change in the $230/kW well cost (see Fig. 24). However, heat exchanger, condenser, and pump costs begin to dominate at P_r's greater than 1.5. Furthermore, heat exchanger and condenser costs at $350/kW represent the major capital investment for the NH_3 cycle while they are below the well cost fraction for the R-32 cycle at the lower temperature. The major cause of this is again increased well capacity (300 lb/sec versus 100 lb/sec) and a higher η_{cycle}. The R-32 cycle at 150°C requires 88 wells at a total capital investment of $89,489,000 whereas the NH_3 cycle at 250°C requires 10 wells at $23,151,000. This difference is even more pronounced when one considers that the specified geothermal gradients were similar (50-60°C/km). The cost of one well in the 250°C resource area was $1,177,800 (at 3930 m) versus $437,500 (at 2500 m) for one in the 150°C resource area.

At this point, we have established that resource temperature and well flow rates are extremely important in controlling the economics of binary-fluid cycles. These facts will be used in the sections that follow as the effect of resource temperature is studied in more detail and a generalized cost model is developed.

Our treatment would not be complete without a comparison between proposed binary-fluid cycles and direct flashing systems such as those currently in use throughout the world, as in the liquid-dominated fields at Cerro Prieto, Mexico and Wairakei, New Zealand. We make a specific comparison using the 150°C resource as specified in Table 14. The major differences between these two types of power conversion systems are observed by examining Tables 15 and 18. The flashing systems do not require a heat exchanger and for this specific case use of a spray condenser would greatly reduce the cost of this component. Properly selected binary cycles would typically have a higher η_u than 2-stage flashing systems (59.2% versus 55.0 and 36.2%) and therefore for the same power

TABLE 18

COST SUMMARY FOR A 100 MW(e) POWER PLANT (1976 DOLLARS)

Equipment	150°C Geothermal Resource			250°C Geothermal Resource
	R-32 Binary-Fluid Cycle ($)	Direct Flashing (T_o = 26.7°C) ($)	Direct Flashing (T_o = 48.9°C) ($)	R-717 (NH_3) Binary-Fluid Cycle ($)
Turbine } Φ_T	308,262	41,723,732	12,559,915	205,200
Generator } Φ_T	1,127,671	1,127,671	1,127.671	1,127,671
Pumps } Φ_P	740,000	{250,000	{250,000	255,000
Turbine drive/or electric } Φ_P	87,574			10,000
Heat exchanger} Φ_{HE}	10,970,278 ($17.5/ft^2)	---	---	3,843,909 ($19/ft^2)
Flashing tanks + cyclones	---	200,000	200,000	---
Condenser/desuperheater} Φ_C	7,276,158 ($9/ft^2)	500,000	500,000	8,769,030 ($15/ft^2)
Total purchased equipment	20,509,943	43,801,403	14,637,586	14,210,810
$(1 + \Sigma f_i)$ direct cost factor = 1.63				
(f_I) indirect cost factor = 1.70				
Equipment capital investment	56,833,052	121,373,688	40,560,751	39,378,155

TABLE 18 (continued)

COST SUMMARY FOR A 100 MW(e) POWER PLANT (1976 DOLLARS)

	150°C Geothermal Resource			250°C Geothermal Resource
	R-32 Binary-Fluid Cycle ($)	Direct Flashing (T_o = 26.7°C) ($)	Direct Flashing (T_o = 48.9°C) ($)	R-717 (NH_3) Binary Fluid Cycle ($)
Wells				
Drilling + casing costs} Φ_W	38,500,000 (88 wells)	42,000,000 (96 wells)	63,437,500 (145 wells)	11,778,000 (10 wells)
$(1 + f_W)$ direct cost factor	1.49	1.50	1.50	1.26
f_I^* indirect cost factor	1.56	1.56	1.56	1.56
Well capital investment	89,489,400	98,280,000	148,443,750	23,150,837
Total capital investment	146,322,452	219,653,688	189,004,501	62,528,992
$/kW installed	1463	2197	1890	625
Annual costs				
Fixed charge rate 17%/yr	24,874,817	37,341,127	32,130,765	10,629,928
Operating	400,000	400,000	400,000	400,000
Maintenance	600,000	600,000	600,000	600,000
Power generating cost at busbar	3.48¢/kWh	5.15¢/kWh[a]	4.45¢/kWh[a]	1.56¢/kWh
85% load factor - 7446 hr/yr				

[a] Actual cost minimum occurs at T_o = 37.8°C with 4.27 ¢/kWh.

output they require fewer wells (88 versus 96 and 145 for 100 MW(e)). The major equipment investment for flashing systems would be the turbines and in many cases condensers and cooling towers. For binary-fluid cycle installations the primary heat exchanger and condenser/desuperheater components comprise the major equipment costs. The results listed in Table 18 illustrate this point. Furthermore, the effect of exhaust temperature on turbine capital investment is given by the difference between $41,723,732 for twenty-five, 2-m-diameter 10-stage exhaust ends with T_0 = 26.7°C versus $12,559,915 for ten, 2-m-diameter 7-stage exhaust ends with T_0 = 49.8°C. This is in contrast to a $308,262 investment for an R-32 turbine of the same 100 MW(e) power capacity. In flashing systems, decreasing turbine cost requirements are matched against increasing well requirements as the exhaust temperature is increased. For this particular 150°C resource, a cost minimum of 4.27¢/kWh actually occurs at T_0 = 37.8°C (100°F) for a 2-stage flashing system; but the cost curve is relatively flat in the region between 37.8 and 49.8°C.

Using optimum conditions for the R-32 cycle, resulting generating costs of 3.48¢/kWh are still less than the best 2-stage flashing case at 4.27¢/kWh. It should be emphasized that any cost estimate depends on the correlations used for each major component, and are subject to change depending on prevailing economic conditions. These costs are both high but represent a continuation of the current level of price escalation that has prevailed for the last few years. Geothermal electric generating costs are compared with present fossil-fuel fired and nuclear generating costs in a later section.

6.8 Resource Temperature Effects on Cycle Economics

As was apparent from the comparison of the 250°C and 150°C cases for the same geothermal gradient, generating costs per kWh were less at the higher temperature. Since well flow rates differed for both cases, we extended the case study of an R-32 binary-fluid cycle to cover a range of geothermal fluid temperatures from 130 to 250°C. At each temperature examined, reduced cycle pressures were varied until an economic optimum was reached. A constant geothermal gradient of 50°C/km with conditions corresponding to those of Table 14 was used. A constant well flow of 45 kg/sec (100 lb/sec) was also assumed. The results, presented in Fig. 25 and Table 19, show that drilling into a 230°C resource produces a minimum generating cost of 3.18¢/kWh. This establishes that there is indeed an optimum depth for a producing geothermal system subject to the constraints of the economic model used.

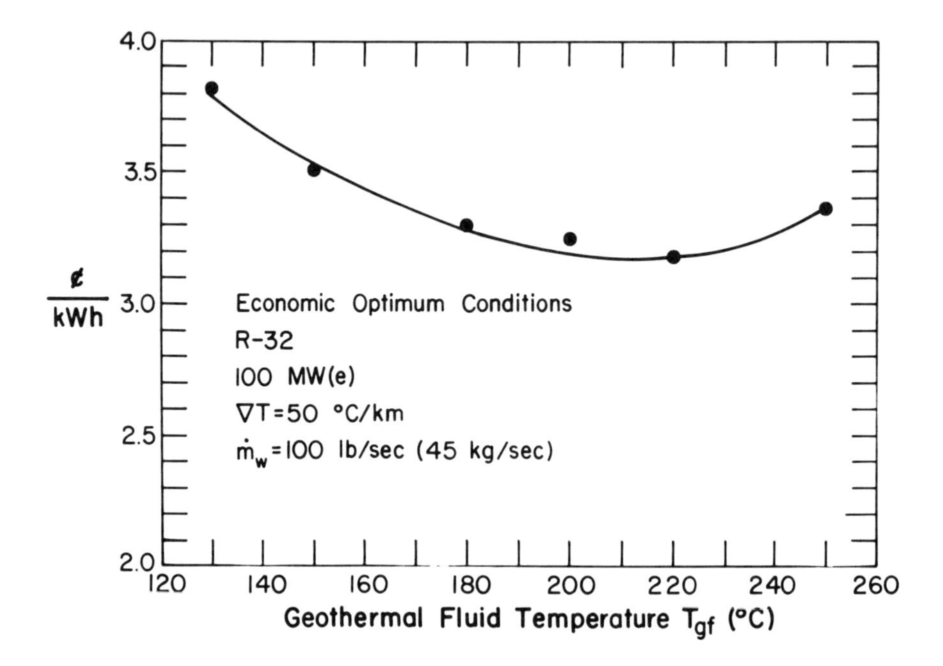

Fig. 25
Generating costs versus geothermal fluid temperature for a R-32 binary-fluid cycle. Well depths correspond to a geothermal gradient of 50°C/km with heat rejection at 26.7°C. Cost estimates based on 1976 dollars.

6.9 Generalized Cost Model

In order to assess the potential for geothermal energy as a viable source of electric power, a generalized approach was developed to aid in the preliminary assessment for a given area where the geothermal gradient is known. Alternate approaches to geothermal power economics are currently being developed at the Battelle Northwest Laboratories[56,57] and Lawrence Berkeley Laboratory[58] and should be of considerable use for estimating purposes. Since the operating cost will depend on the number and depth of required wells and on the equipment cost associated with the surface conversion plant, a simplified cost model was created in which the installed generating cost could be expressed as a function of the geothermal fluid temperature (T_{gf}), geothermal gradient (∇T), and well flow rate ($\dot{m}_w$). The model assumes a two-hole circulating, dry hot rock system with an equal number of production and reinjection wells. A net 100 MW(e) binary-fluid cycle was selected because it potentially represents an efficient conversion process for producing electricity. Equipment costs Φ_E were calculated for 100 to 300°C geothermal resource temperatures and several working fluids. Over this temperature range Φ_E varied as a linearly decreasing function of temperature:

TABLE 19

ECONOMIC OPTIMUM CONDITIONS FOR A 100 MW(e) R-32 CYCLE (1976 DOLLARS)

$\dot{m}_w$ = 45 kg/sec (100 lb/sec) ∇T = 50°C/km

Geothermal Fluid Temperature T_{gf} (°C)	Well Depth (m)	Number of Wells[a]	η_u (%)	η_{cycle} (%)	Reduced Cycle Pressure P_r	$/kW Installed	¢/kWh
130	2100	132	54.4	12.0	1.25	1618	3.83
150	2500	88	59.2	12.9	1.25	1463	3.48
180	3100	54	65.1	15.4	1.78	1388	3.30
200	3500	44	63.7	15.9	1.78	1378	3.28
230	4100	35	61.3	16.5	1.78	1333	3.18
250	4500	29	64.3	18.6	2.55	1420	3.38

[a] Includes production and reinjection wells.

$$\Phi_E(\$/kW) = 320.5 - 0.7040\, T_{gf} \qquad (67)$$

where T_{gf} is expressed in °C. In addition, individual well costs Φ^*_W could be approximated by an exponential function in depth Z using the estimate given in Fig. 17:

$$\Phi^*_W(\$/\text{well}) = \left(64.3949 \exp\left[3.88843 \times 10^{-4} Z\right]\right) Z \qquad (68)$$

where Z is expressed in meters. The depth can also be expressed as a function of the geothermal gradient and the ambient crustal temperature (T^*_o):

$$Z = \frac{T_{gf} - T^*_o}{\nabla T} \simeq \frac{1000\left(T_{gf} - 15°C\right)}{\nabla T} \qquad (69)$$

where T_{gf} is expressed in °C and ∇T in °C/km. The geothermal fluid flow rate is equivalent to the net power output divided by η_u for the fluid of interest times ΔB. Therefore, the number of wells is given by:

$$n_W = \frac{2P}{\eta_u \Delta B \dot{m}_W} = \frac{200 \times 10^6}{\eta_u \Delta B \dot{m}_W} \text{ (for 100 MW(e) plant)} \qquad (70)$$

where the factor of 2 accounts for production and reinjection wells and $\dot{m}_W$ is expressed in kg/sec and ΔB in J/kg. Using a mean value of 60% for η_u (see Table 19) and approximating ΔB by Eq. (19) with C_p = 4200 J/kgK, the total well cost Φ_W can be expressed as:

$$\Phi_W(\$/kW) = \frac{n_W \Phi^*_W}{100{,}000}$$

$$\Phi_W(\$/kW) = \frac{51.107\, Z \exp\left[3.8884 \times 10^{-4} Z\right]}{\dot{m}_W\left[T_{gf} - T_o - \left(T_o + 273.15\right)\ln\left(\frac{T_{gf} + 273.15}{T_o + 273.15}\right)\right]} \qquad (71)$$

Using average values for the direct and indirect cost factors (Tables 12 and 13):

$$1 + f_W = 1.5 \qquad f_I = 1.70$$

$$1 + \Sigma f_i = 1.63 \qquad f^*_I = 1.56 \qquad (72)$$

and assuming $T_0 = 26.7°C$, we can combine Eqs. (67) and (71) to give the total installed cost per kW:

$$\Phi(\$/kW) = \frac{119.59\ Z \exp\left[3.8884 \times 10^{-4}\ Z\right]}{\dot{m}_w\left[T_{gf} - 26.7 - 299.85\ \ln\left(\frac{T_{gf} + 273.15}{299.85}\right)\right]}$$

$$+ 888.1 - 1.951\ T_{gf}\ . \qquad (73)$$

Equation (73) is applicable for geothermal fluid temperatures ranging from 100 to 300°C (212 to 572°F). At temperatures above 300°C, the installed equipment cost levels off at approximately $277/kW, but due to the uncertainty of anticipated geochemical problems we suspect that the equipment cost might escalate significantly.

Equation (73) was modified for estimating generating costs in ¢/kWh. An 85% (7446 h/yr) load factor and a 17% annual fixed charge rate were assumed. An additional 0.13 ¢/kWh should be added to cover annual operating and maintenance costs. Generalized charts, calculated using these assumptions and Eq. (73), are presented in Fig. 26 (parts A, B, and C). Three well-flow rates, 45, 113, 227 kg/sec (100, 250, and 500 lb/sec), and geothermal gradients ranging from 20 to 200°C/km were investigated for geothermal temperatures from 100 to 300°C. At the lower gradients, 20 to 50°C/km, distinct minimums in the cost curves are seen before T_{gf} reaches 300°C. For the poorest gradient of 20°C/km, these minima are actually at temperatures slightly less than 100°C depending on $\dot{m}_w$. When T_{gf} drops much below 80°C, the cycle efficiency is so low that the costs begin to rise rapidly again. For the larger gradients, 70°C/km and above, cost minima are not even reached at 300°C. At first glance, the implication of this might be that one should keep on drilling to deeper depths and higher temperatures; but the equipment costs are leveling off at temperatures just above 300°C and this effect causes the costs to again increase before temperatures reach 320°C. Nonetheless, the curves do indicate that when utilizing a geothermal resource having a known gradient serious attention should be given to where that minimum cost may occur. Although the actual position of the minimum is subject to the constraints and assumptions of the cost model, one can infer that it will occur at geothermal temperatures below 200°C as the gradient falls off to 20°C/km.

In order to assess the geothermal potential of the entire country, the results of two summaries of geothermal heat flow data[59,60] were transformed into gradients using acceptable thermal conductivities. The map shown in Fig. 27 provides contour lines of constant geothermal gradient. Several anomalous areas with higher gradients are also located;

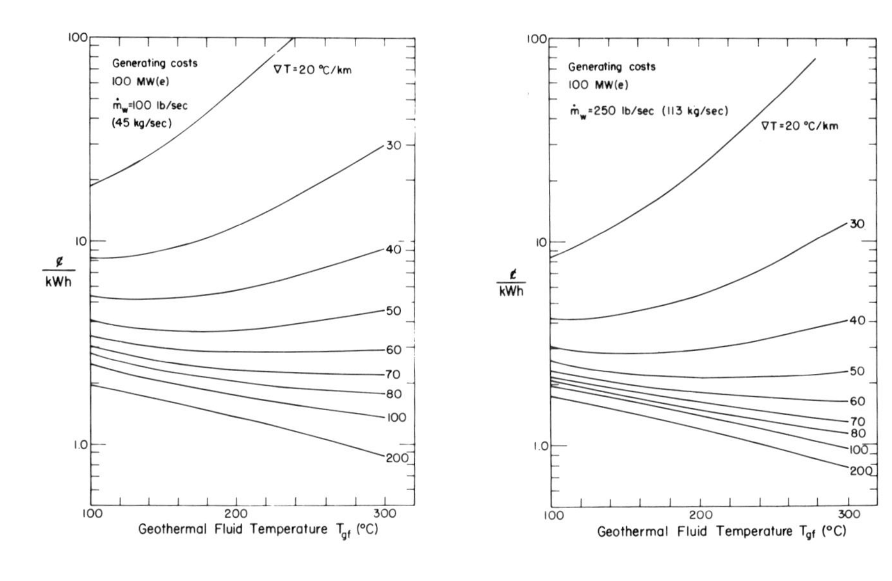

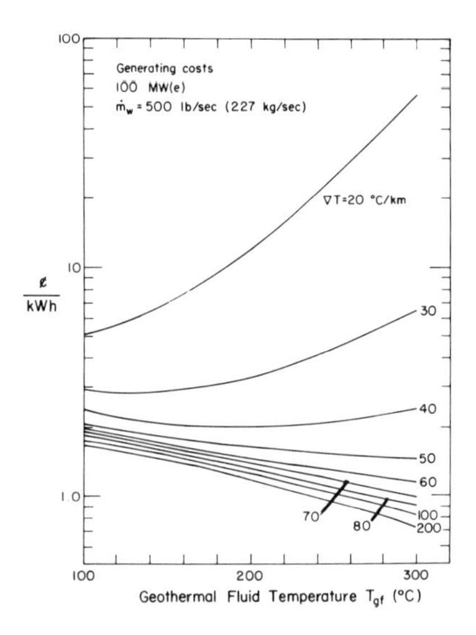

Fig. 26

Generalized cost model for geothermal systems. Generating costs expressed as a function of well flow rate $\dot{m}_w$, geothermal gradient ∇T, and geothermal fluid temperature T_{gf}. Cost estimates based on 1976 dollars.

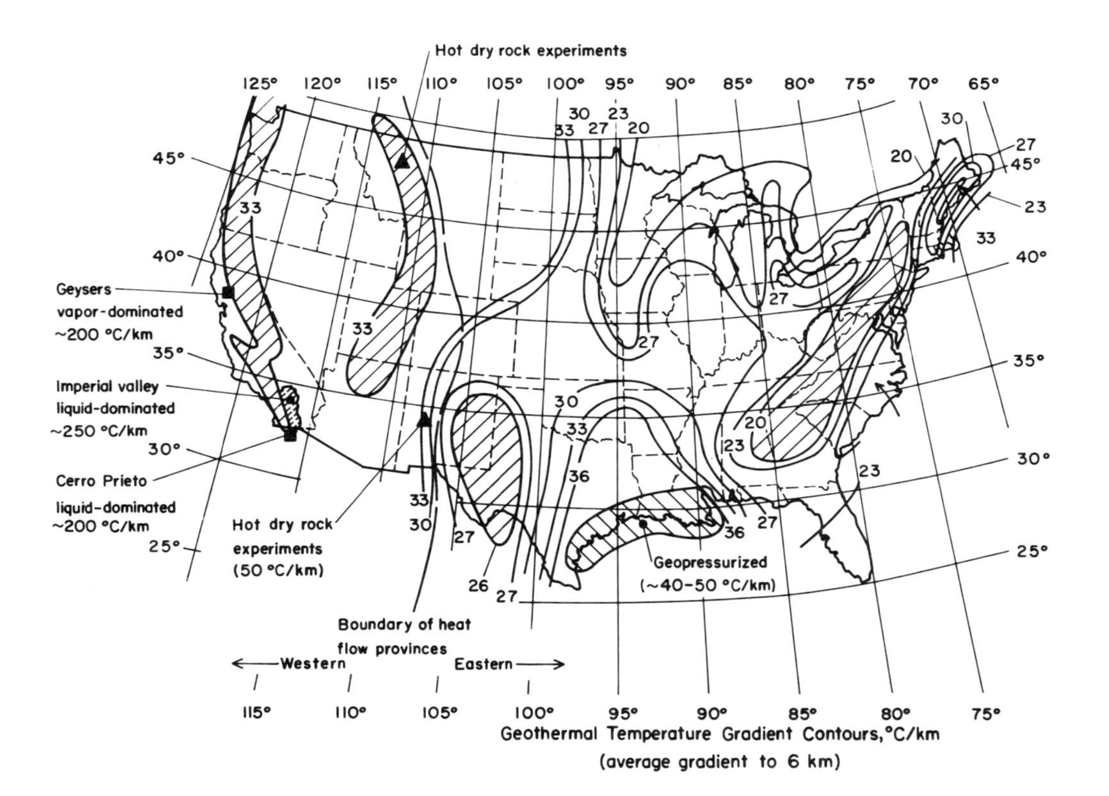

Fig. 27

Map of United States showing lines of constant average geothermal gradient.

but the map itself is almost anomaly free and probably conservative. Using these gradients and the generalized charts of Fig. 26, one can determine approximate generating costs.

One important factor is the range of gradients from 20 to 36°C/km that might be found in nonanomalous locations. In terms of minimum generating costs for a 100 MW(e) plant, a 20°C/km gradient corresponds to a 5¢/kWh busbar cost at a high-well-flow rate of 227 kg/sec (500 lb/sec). A 40°C/km gradient at the same well-flow rate corresponds to approximately a cost of 2¢/kWh. The costs, of course, escalate rapidly as the well-flow rate decreases.

Before leaving this particular topic several factors should be emphasized. Economic factors for geothermal power development tend to be very site specific and each case should be treated separately. Furthermore, our entire economic analysis depends largely on currently available technology and its associated cost factors. This includes downhole drilling technology as well as surface plant equipment. Engineering advances will no doubt occur as the country's geothermal program matures, so our economic estimates may be expected to be conservative. In particular, if concepts such as the total flow or the helical two-phase expander are developed one might expect a significant reduction in capital equipment investments.[9]

7. GEOTHERMAL ENERGY AS A COMPETITIVE PRODUCER OF POWER — CONCLUDING REMARKS

Of the 380,000 MW of electricity currently produced in the United States only 390 MW are produced by geothermal energy.[8] Clearly there is a long road ahead if geothermal power is to have a significant impact in the next thirty years. One major factor that will influence the rate of geothermal development is obviously the current cost of more conventional energy sources such as oil, gas, coal, and nuclear. Numerous books and papers have addressed part or all of this topic in recent years [61-66] and what we have attempted to do is update their findings,particularly in view of the current increase in fossil-fuel costs. Recent studies by Meidav [61,62] and Bupp et al.[63] indicate that fossil-fuel costs alone have escalated to an equivalent 2¢/kWh for oil and 1¢/kWh for coal. Nuclear fuel costs remain in the 0.24 to 0.30¢/kWh range. Construction and equipment costs have also greatly increased in the past two years. In many cases, environmental controls have escalated costs and will continue to do so. A breakdown of equipment and fuel cost ranges is presented in Table 20 for geothermal, fossil-fuel, and nuclear electric power production plants. Generating costs for flashing plants (1.61 to 4.30¢/kWh) and for binary-fluid cycle plants (1.56 to 4.18¢/kWh) are within the ranges for nuclear or fossil-fuel sources. Geothermal power plant costs are very capital intensive, much like a hydroelectric plant, and this does create problems in attracting public utility investment in a geothermal system, particularly when interest rates are high. However, as oil, gas, and coal supplies become more difficult and costly to obtain, geothermal power could play a vital role in our energy-producing economy. Improved drilling techniques and enhancement of overall heat transfer coefficients would make geothermal systems even more attractive.

As mentioned earlier, in calculating the total cost of electricity, costs should be added for the transmission and distribution system required between the power plant busbar and the customer. Transmission networks can be expensive particularly if the power-generating facility is remotely located. If geothermal power plants are limited to only natural systems, in many cases the sites would be remote and the transmission costs high. Dry hot rock systems should not be overly restricted in their locations and therefore transmission costs will probably be competitive with fossil-fuel or nuclear generating systems. Environmental concerns frequently restrict the location of nuclear plants and to some extent fossil-fuel fired plants.

In cases where the geothermal gradient is low, e.g. 20°C/km, a hybrid system might be effective. In such systems only part of the geothermal heat is converted to electricity and the remainder rejected at a high enough temperature to be useful for direct applications such as space heating or industrial process heat. In many cases, direct (nonelectrical) applications for geothermal energy are practical, but again the economics of transporting the geofluid itself can be critical in determining whether the concept is feasible. For further information in this area the reader is referred to the extensive work by Beall et al. at Oak Ridge National Laboratory[67-69] and Bodvarsson, Reistad, et al.[70] who have examined direct applications for a low-grade heat source such as geothermal.

The major purpose of this monograph is to establish the engineering criteria for electric power production from geothermal resources. Thermodynamic and economic guidelines were developed for evaluating specific working fluids as well as for direct-steam flashing systems. Since well costs comprise a significant fraction of the capital investment for a power plant, a utilization efficiency (η_u) was used in evaluating potential working fluids. The specific characteristics of any geothermal resource, in particular its temperature, depth from the surface, and well-flow capacity, need to be specified before an optimum working fluid can be selected. There are no simple answers and a variety of alternatives should be considered for each case.

TABLE 20

COMPARISON OF FOSSIL-FUEL, NUCLEAR, AND GEOTHERMAL ESTIMATED GENERATING COSTS (1976 DOLLARS)

Resource Type	Installed Equipment Costs ($/kW)	Equipment Cost as (¢/kWh)[a]	Operating and Maintenance as (¢/kWh)	Well or Fuel Cost (¢/kWh)	Total Generating Cost (¢/kWh)
Direct flashing[b]	300-600	0.68-1.37	0.13	0.80-2.80	1.61-4.30
Binary-fluid cycles[b]	400-700	0.91-1.60	0.13	0.53-2.45	1.57-4.18
Nuclear[d]	$\geq$800	$\geq$1.83	0.13	0.30	$\geq$2.26
Fossil fuel - oil[e]	400-600[c]	0.91-1.37	0.13	2.0 ($12/bbl)	3.04-3.50
Fossil fuel - coal[e]	400-600[c]	0.91-1.37	0.13	1.0 ($25/ton)	2.04-2.50

[a] 17% annual fixed charge rate.
85% (7446 hr/yr) load factor.

[b] 150-200°C resources. Plant startup in 1980.
$\dot{m}_w$ = 100-300 lb/sec.

[c] Higher costs correspond to stringent environmental control systems.[63,64,65]

[d] Plant startup in 1984, see Ref. 62-65 for cost estimates.

[e] Plant startup in 1980, see Ref. 62-65 for cost estimates.

APPENDIX A

Direct Steam Flashing

A direct flashing system differs from a binary-fluid system in that part of the geothermal fluid is itself used as the thermodynamic working medium thus eliminating the primary heat exchanger and feed pump. Because of the inherent simplicity of this system, performance evaluations can be made with relative ease; for instance, for one and two stages of flashing, it is possible to estimate analytically the conditions of temperature and pressure at each flashing step for optimum thermodynamic performance.

Generally, the work developed by a given flash stage is proportional to the amount of vapor created by the throttling step (isenthalpic expansion) and the resultant enthalpy difference that the vapor would experience in an isentropic expansion to the cycle heat rejection temperature. If the temperature to which the fluid is flashed is only slightly below the wellhead temperature, then the fraction of vapor produced will be small but the isentropic enthalpy difference will be large. The opposite is true if the fluid is flashed to just above the heat rejection temperature. Maximum work output will result at some intermediate value.

Assuming that the geothermal fluid can be characterized by saturated liquid (water) at temperature ($T_{gf} = T$) and specific enthalpy $H_\ell(T)$ (saturation values are implied), then an isenthalpic expansion to a lower temperature T_1 will divide the initial flow ($\dot{m}$) into vapor and liquid components given respectively by Eqs. (A-1) and (A-2).

$$\dot{m}_{g1} = \frac{\dot{m}\left[H_\ell\left(T\right) - H_\ell\left(T_1\right)\right]}{h_{fg}\left(T_1\right)} \tag{A-1}$$

$$\dot{m}_{\ell 1} = \dot{m}\left[1 - \frac{\left[H_\ell\left(T\right) - H_\ell\left(T_1\right)\right]}{h_{fg}\left(T_1\right)}\right] \tag{A-2}$$

In the usual manner, the liquid and vapor fractions are separated and power from the first flash stage is obtained by expanding the vapor in a turbine to the temperature T_0 where heat is rejected from the cycle The amount of power developed is equal to the product of the vapor mass flow as given by Eq. (A-1) and the specific enthalpy drop of the vapor during the turbine expansion,

$$P_1 = \frac{\dot{m}\left[H_\ell(T) - H_\ell(T_1)\right]}{h_{fg}(T_1)} \eta_t \left\{\left[H_g(T_1) - H_g(T_o)\right] - T_o\left[S_g(T_1) - S_g(T_o)\right]\right\} , \quad \text{(A-3)}$$

where η_t represents an average turbine efficiency that accounts for the presence of moisture in the turbine expansion ($\eta_t \simeq 0.80$).

If a dual stage flash is employed, the residual liquid remaining from the first flash is in turn flashed to a lower temperature, T_2, and the resulting vapor is expanded in a turbine to provide additional power. The expression for the second stage power in this cascading process is obtained by replacing T_1, T, and $\dot{m}$ in Eq. (A-3) by the quantities T_2, T_1, and $\dot{m}_{\ell 1}$. The result is

$$P_2 = \eta_t \dot{m}\left[1 - \frac{\left[H_\ell(T) - H_\ell(T_1)\right]}{h_{fg}(T_1)}\right]\left[\frac{\left[H_\ell(T_1) - H_\ell(T_2)\right]}{h_{fg}(T_2)}\right]\left\{H_g(T_2) - H_g(T_o) - T_o\left[S_g(T_2) - S_g(T_o)\right]\right\} . \quad \text{(A-4)}$$

Optimum thermodynamic performance occurs at those values of T_1 and T_2 for which both partial derivatives of the expression for the combined power vanish. To this end, a number of simplifications are possible which not only facilitate the differentiation process but also allow for an analytical solution. Without any loss in generality, the properties of the saturated vapor may be expressed in terms of the saturated liquid properties by the relations

$$H_g(T) = H_\ell(T) + h_{fg}(T) , \quad \text{(A-5)}$$

and

$$S_g(T) = S_\ell(T) + \frac{h_{fg}(T)}{T} . \quad \text{(A-6)}$$

Using the thermodynamic identity TdS = dH − v dP and assuming that the specific heat of the saturated liquid is constant (i.e., $\Delta H_\ell = C^{sat}_{p\ell}\Delta T$) the expression for the first stage power will take the form

$$P_1 = \frac{\dot{m}C^{sat}_{p\ell}\left(T - T_1\right)\eta_t}{h_{fg}\left(T_1\right)} \left\{ C^{sat}_{p\ell}\left(T_1 - T_o\right) - T_o C^{sat}_{p\ell} \ln\left(T_1/T_o\right) + h_{fg}\left(T_1\right)\left(1 - \frac{T_o}{T_1}\right)\right\} \quad \text{(A-7)}$$

where the effect of pressure on the entropy of the liquid has been neglected (because v is small). By expanding the logarithmic term in powers of $(T_1 - T_o)/T_o$ and keeping terms only up to second order, Eq. (A-7) reduces to

$$P_1 = \frac{\dot{m}C^{sat}_{p\ell}\left(T - T_1\right)\eta_t T_o}{h_{fg}\left(T_1\right)} \left\{ \frac{C^{sat}_{p\ell}}{2}\left(\frac{T_1 - T_o}{T_o}\right)^2 + \left(\frac{T_1 - T_o}{T_1 T_o}\right) h_{fg}\left(T_1\right)\right\} . \quad \text{(A-8)}$$

For flashing temperatures not exceeding 200°C, the contribution from the first term within brackets can be neglected without much loss in accuracy, and the first stage power reduces to the following simple result

$$P_1 = \eta_t \dot{m}C^{sat}_{p\ell}\left(T - T_1\right)\left(\frac{T_1 - T_o}{T_1}\right) . \quad \text{(A-9)}$$

If only one stage of flashing is being considered, then Eq. (A-9), which represents the total power developed, will be a maximum when the fluid is flashed to the value of T_1 given by

$$T_1^{opt} \text{ (single stage flash)} = \sqrt{T_o T} \quad \text{(A-10)}$$

which is the geometric mean of T_o and T, not the arithmetic mean.

If two stages of flashing are used, then the power from the second stage flash is as follows

$$P_2 = \dot{m}C^{sat}_{p\ell}\,\eta_t\left(1 - \frac{C^{sat}_{p\ell}\left(T - T_1\right)}{h_{fg}\left(T_1\right)}\right)\left(T_1 - T_2\right)\left(\frac{T_2 - T_o}{T_2}\right) \quad \text{(A-11)}$$

Neglecting variations in latent heat, optimum performance for the combined system will occur at values of T_1 and T_2 that are solutions to the following set of algebraic equations:

$$\frac{\partial\left(P_1 + P_2\right)}{\partial T_1} = 0 => \frac{T_o T}{T_1^2} - \frac{T_o}{T_2} - \frac{C_{p\ell}^{sat}}{h_{fg}\left(T_1\right)}\left(\frac{T_2 - T_o}{T_2}\right)\left(T - 2T_1 + T_2\right) = 0 \ . \tag{A-12}$$

$$\frac{\partial\left(P_1 + P_2\right)}{\partial T_2} = 0 => T_2^2 - T_o T_1 = 0 \ . \tag{A-13}$$

In accordance with the result of Eq. (A-10), the solution to Eq. (A-13) is simply

$$T_2^{opt} = \sqrt{T_o T_1^{opt}} \ . \tag{A-14}$$

In Eq. (A-12) only a very small error is made if the terms involving $C_{p\ell}^{sat}/h_{fg}$ are neglected. This establishes an optimum T_1 value of

$$T_1^{opt} = \sqrt{T T_2^{opt}} = \left(T_o T^2\right)^{1/3} \ . \tag{A-15}$$

For any given combination of T and T_o, the optimum flash temperatures can be determined quickly, and Eqs. (A-3) and (A-4) can be used along with the steam tables to establish the total optimized power for the two stage system. If only a single flashing step is being considered, then Eqs. (A-10) and (A-3) should be used.

APPENDIX B

Heat Exchanger Optimization

From the standpoint of economical power generation, the optimum allocation of the available system temperature drop between the various parasitic and productive thermodynamic cycle components (heat exchangers and pumps in the former and turbine in the latter) is in large part dependent upon the relative costs of obtaining the geothermal fluid, transporting its heat to the system, and rejecting the waste heat to the environment. If drilling costs are low, then presumably a larger fraction of the resource to environment temperature difference can be allocated to the heat transfer steps in order to minimize the size and hence cost of heat exchange equipment. On the other hand, if the cost of obtaining the geothermal fluid is high in comparison with the cost of heat exchange equipment, then the incentive is strong for minimizing the potential work (availability) which is lost in the heat exchange steps. Thus, a smaller fraction of the available system temperature drop would be consumed in these steps.

In general, the appropriate balance between heat exchanger and well costs is difficult to ascertain since this implies making numerous thermodynamic cycle calculations and interfacing the results with component cost estimates. The approximate magnitude of the temperature difference between geothermal and working fluid streams (as well as between the working fluid and power plant coolant streams) can be estimated by considering a simplified thermodynamic model which eliminates the complexities of working fluid property variations but retains the desirable characteristics of providing a high well utilization factor η_u.

For prescribed values of approach and pinch point temperature differences in the primary heat exchanger, maximum resource utilization frequently occurs at supercritical operating pressures where the specific heat of the working fluid does not change appreciably throughout the heating process. (See Chapter 4) Consequently, one can envision a hypothetical idealized, binary-fluid cycle corresponding to the temperature-entropy diagram of Fig. B-1. To simplify the discussion, a uniform temperature difference ΔT_1 is prescribed in the primary heat exchanger, and the amount of heating of the power plant coolant in the heat rejection system is assumed small. Under these conditions the cycle rejects heat at ΔT_2 degrees above the sink temperature T_o. In this example, the turbine and pump are assumed to be ideal and consequently irreversibilities are introduced only in heat exchange steps where there are net increases in entropy.

When the finite irreversibilities of the pump and turbine are considered, then only a fraction of the work obtained by the ideal cycle of Fig. B-1 can be realized. Calling this fraction ϵ, and assuming that

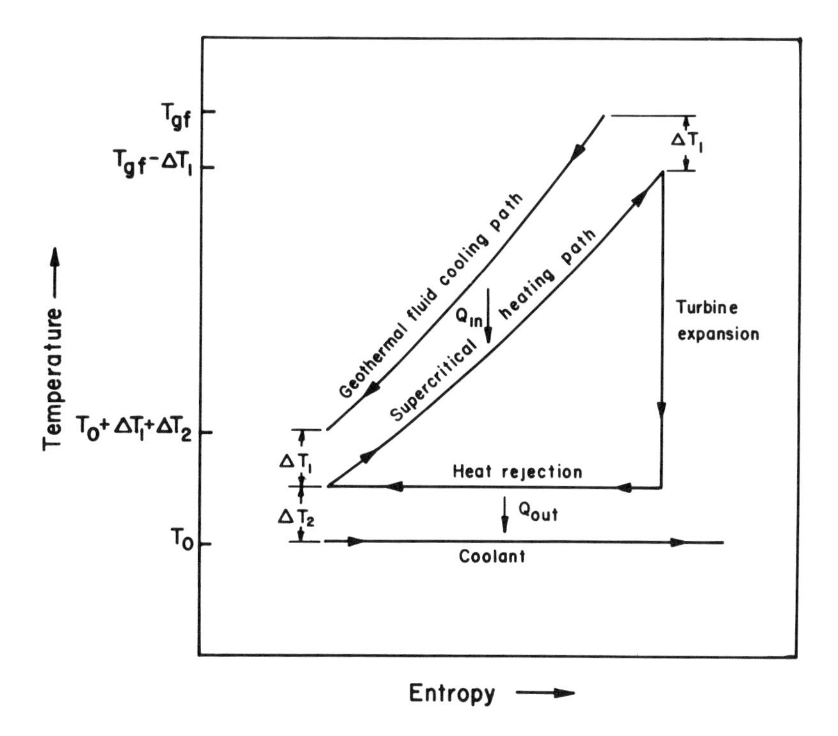

Fig. B-1
Temperature-entropy diagram for an idealized thermodynamic cycle.

temperature differences across both heat exchange surfaces are uniform, the power developed from a real (but still mathematically simplified) cycle is given by

$$P = \varepsilon \dot{m}_{gf} C_{p_\ell} \left\{ \left(T_{gf} - \Delta T_1\right) - \left(T_o + \Delta T_2\right) \left[1 + \ell n \frac{\left(T_{gf} - \Delta T_1\right)}{\left(T_o + \Delta T_2\right)} \right] \right\}$$

(B-1)

where T_{gf} is taken as the wellhead temperature of the geothermal fluid and $\dot{m}_{gf}$ its mass flow rate. As before, the specific heat $C_{p\ell}$ of the fluid is assumed constant.

When ϵ attains its limiting value of unity, Eq. (B-1) reduces to the expression for the net maximum useful work attainable from a liquid geothermal fluid stream whose source temperature has been reduced by an amount ΔT_1 and whose ultimate heat rejection temperature is ΔT_2 degrees above the sink temperature T_o (compare with Eq. (19)). Consequently, for a given net power, an increase in the heat transfer driving force (ΔT_1 or ΔT_2) results in an increase in well flow ($\dot{m}_{gf}$). Optimum values of ΔT_1 and ΔT_2 are those which balance the cost of heat exchange equipment with the cost of obtaining the heat.

The total amount of surface area required for heat transfer is proportional to the cost of heat exchange equipment, and can be determined from the amounts of heat transferred to the cycle and rejected to the environment, the temperature differences ΔT_1 and ΔT_2, and the average heat transfer coefficients U_1 and U_2 associated with the primary heat exchanger and heat rejection system respectively. Calling the primary heat exchange area A_1 and the condenser area A_2, the defining relationships are

$$A_1 = \frac{Q_{in}}{U_1 \Delta T_1} \qquad \text{(B-2)}$$

$$A_2 = \frac{Q_{out}}{U_2 \Delta T_2} \qquad \text{(B-3)}$$

where

$$Q_{in} = \dot{m}_{gf} C_{p_\ell} \left(T_{gf} - T_o - \Delta T_1 - \Delta T_2 \right) \qquad \text{(B-4)}$$

and

$$Q_{out} = Q_{in} - P \qquad \text{(B-5)}$$

Equations (B-1)-(B-5) provide the basis from which a cost optimization for the simplified model may be obtained. Assuming overall plant cost is represented by the sum of its component parts, a general expression can be written which relates the total capital investment to the unit cost of heat exchangers, wells and remaining equipment and structures. Following the method used in Chapter 6 the total cost Φ is given by,

$$\Phi = C_1 A_1 + C_2 A_2 + C_{gf} \dot{m}_{gf} + \text{Cost Remainder} \qquad \text{(B-6)}$$

where $\mathcal{C}_1$ and $\mathcal{C}_2$ are the cost per unit surface area of the primary heat exchanger and heat rejection systems respectively and $\mathcal{C}_{gf}$ the cost per unit of well flow rate which reflects the drilling cost of production and reinjection wells and piping to the central power plant. For plants larger than demonstration size the remainder of the plant cost will be proportional to the net generating capacity. Then by using Eqs. (B-2)-(B-5) to substitute for the quantities A_1, A_2, $\dot{m}_{gf}$ and by introducing the dimensionless variables

$$\eta = 1 - \frac{\Delta T_1}{T_{gf}}, \quad \xi = 1 + \frac{\Delta T_2}{T_o}$$

Φ becomes

$$\Phi = \frac{P\left\{\frac{\mathcal{C}_1}{U_1 \varepsilon T_{gf}} \frac{\left(T_{gf}\eta - T_o\xi\right)}{(1-\eta)} + \frac{\mathcal{C}_2}{U_2 \varepsilon T_o} \frac{\left(T_{gf}\eta - T_o\xi\right)}{(\xi - 1)} + \frac{\mathcal{C}_{gf}}{C_{p_\ell}\varepsilon}\right\}}{T_{gf}\eta - T_o\xi\left[1 + \ell n\left(\eta/\xi\right) + \ell n\left(T_{gf}/T_o\right)\right]}$$

$$- P\frac{\mathcal{C}_2}{U_2 T_o\left(\xi - 1\right)} + \mathcal{C}_{remainder}P \qquad \text{(B-7)}$$

Because the last term is independent of the variables ξ and η, it need not be considered in calculating the location of the cost minimum.

The condition at the minimum then requires that the partial cost due to wells and heat exchangers be stationary with respect to the variables ξ and η, or

$$\nabla\left\{\frac{\frac{\mathcal{C}_1}{U_1 \mathcal{C}_{gf}\varepsilon} \frac{\left(T_{gf}\eta - T_o\xi\right)}{T_{gf}\left(1 - \eta\right)} + \frac{\mathcal{C}_2}{U_2 \mathcal{C}_{gf}\varepsilon} \frac{\left(T_{gf}\eta - T_o\xi\right)}{T_o(\xi - 1)} + \frac{1}{C_{p_\ell}\varepsilon}}{T_{gf}\eta - T_o\xi\left[1 + \ell n\left(\eta/\xi\right) + \ell n\left(T_{gf}/T_o\right)\right]} - \frac{\mathcal{C}_2}{U_2 \mathcal{C}_{gf} T_o(\xi - 1)}\right\} = 0 \qquad \text{(B-8)}$$

Clearly, the values of η and ξ that are solutions to Eq. (B-8) are functions only of the geothermal source temperature T_{gf}, the heat sink temperature T_o, and the parameters ϵ, $\mathcal{C}_1/U_1\mathcal{C}_{gf}$, and $\mathcal{C}_2/U_2\mathcal{C}_{gf}$. For prescribed values of these quantities, a stationary point may be located by starting from assumed values of $\eta<1$ and $\xi>1$ and proceeding in a stepwise fashion along the direction in the η-ξ plane that is parallel to the local negative gradient vector until a point is reached where the derivatives of the partial cost vanish. For a particular set of conditions

T_{gf}, T_o and ϵ this procedure may be repeated several times for different values of $\mathcal{C}_1/U_1\mathcal{C}_{gf}$ and $\mathcal{C}_2/U_2\mathcal{C}_{gf}$ until a parametric map of optimized values of ξ, η and hence ΔT_1 and ΔT_2 is obtained.

The results of a typical calculation are presented in Fig. B-2 for the conditions T_{gf} = 150°C (300°F), T_o = 26.7°C (80°F) and ϵ = 0.8 (indicating that the thermodynamic cycle operates at 80% of its ideal or limiting value). Shown are the optimum values of ΔT_1 and ΔT_2 for a wide range of values of $\mathcal{C}_1/U_1\mathcal{C}_{gf}$ and $\mathcal{C}_2/U_2\mathcal{C}_{gf}$. Several interesting points are apparent. First, if either or both of the heat transfer coefficients U_1 and U_2 are high and/or the relative cost of heat transfer equipment is small (that is $\mathcal{C}_1/U_1\mathcal{C}_{gf}$ or $\mathcal{C}_2/U_2\mathcal{C}_{gf}$ are small) then a smaller fraction of the overall system temperature drop ($T_{gf}-T_o$) is allocated to the non-productive (in a power generation sense) heat transfer steps. On the other hand, as the relative cost of heat exchange equipment increases and/or the heat transfer coefficients become poor (that is $\mathcal{C}_1/U_1\mathcal{C}_{gf}$ or $\mathcal{C}_2/U_2\mathcal{C}_{gf}$ are

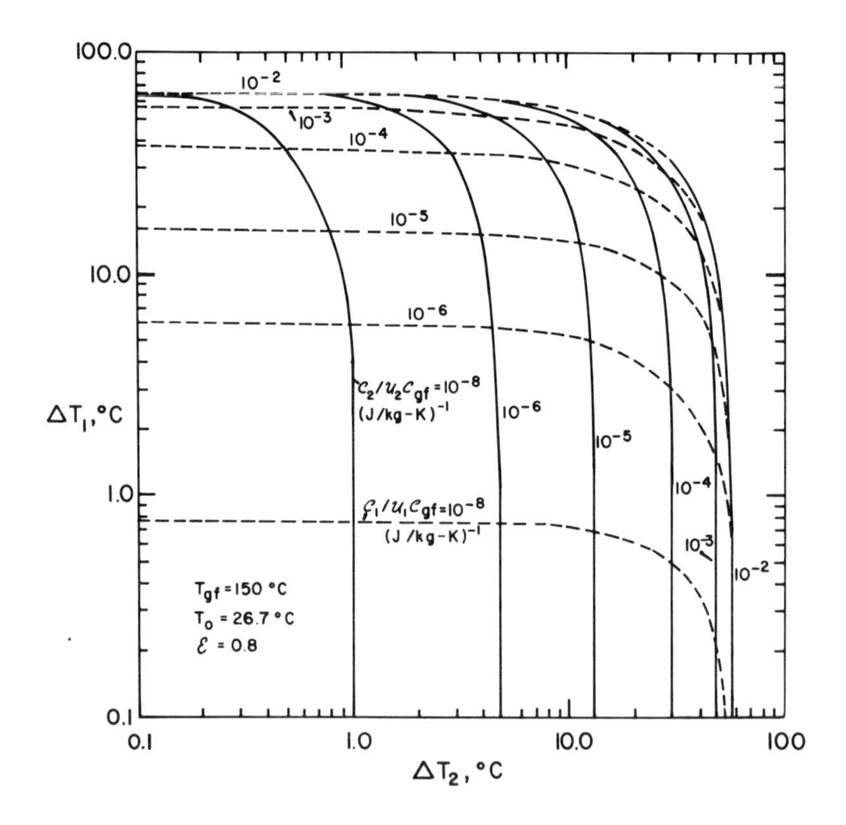

Fig. B-2
Cost–optimized heat exchanger temperature differences for 150°C geothermal wellhead temperature showing the relative effects of heat exchanger unit costs.

large) then larger temperature differences are needed in order to minimize cost. There is, however, an upper limit to the fraction of the overall temperature drop which can be allocated to the heat transfer steps beyond which larger temperature differences result in increased cost. This upper bound is reached when the cost of obtaining the geothermal fluid is small in comparison to the unit heat exchanger costs. For this example, this maximum total allowable ΔT is about 65°C and is represented by the bounding curves $\mathcal{C}_1/U_1\mathcal{C}_{gf} = \mathcal{C}_2/U_2\mathcal{C}_{gf} = 1\times10^{-2}$ $(J/kgK)^{-1}$. Additional curves for parameters higher than this value are crowded in the same region and hence give nearly the same result as the 10^{-2} curves. For typical values of $U_1 \simeq 1000\ W/m^2K$ (170 $BTU/hrft^2°F$) and $\mathcal{C}_1 \simeq \$200/m^2$, the 10^{-2} curves would correspond to a value of $\mathcal{C}_{gf}$ of only \$20/kg/sec which reflects drilling costs far below those anticipated (see Section 6.2). For all practical purposes, the 10^{-2} curves represent an upper limit for geothermal applications.

The existence of an upper limit in heat exchanger temperature differences is not surprising, since for systems for which the dominant cost is attributable to heat transfer equipment, a point is reached at which an increased ΔT results in a correspondingly larger increase in the heat, Q, that must be transferred to maintain a given power level. This is because for large temperature differences the thermodynamic cycle efficiency is poor, accordingly the required heat transfer area and hence cost begin to increase [see Eqs. (B-2) and (B-3)]. The point is simply that one cannot choose the heat exchanger temperature differences arbitrarily large and expect to minimize equipment costs.

Finally, this simplified model estimates heat exchanger temperature differences for representative values of the relative cost factors and mean heat transfer coefficients. Taking typical values as, $\mathcal{C}_1 \simeq \$200/m^2$, $U_1 \simeq 1000\ W/m^2K$ (170 $BTU/hrft^2°F$), $\mathcal{C}_2 \simeq \$150/m^2$, $U_2 \simeq 750\ W/m^2K$ (130 $BTU/hrft^2°F$), and $\mathcal{C}_{gf} \simeq \$4{,}000/kg/sec$ (\$200,000 per well at 50 kg/sec) then both $\mathcal{C}_1/U_1\mathcal{C}_{gf}$ and $\mathcal{C}_2/U_2\mathcal{C}_{gf}$ are found to have the value 5×10^{-5} $(J/kgK)^{-1}$. From Fig. B-2 this results in values for ΔT_1 and ΔT_2 approximately equal to 18°C and 16°C respectively. This particular value of ΔT_1 is within the range of values (13-26°C) used earlier in the thermodynamic cycle calculations of Chapter 4 and economic optimizations of Chapter 6.

APPENDIX C

Economic Factors for Geothermal Reservoirs

C.1 Well Costs

The curves presented in Fig. 17 were derived from existing cost data on geothermal and oil and gas wells. The oil and gas well data include thousands of individual well costs for producing and dry wells throughout the U.S. (continent, offshore, and Alaska). Annual summaries of these costs are published by the Joint Association Survey of the American Petroleum Institute, Independent Petroleum Association of America and the Mid-Continent Oil and Gas Association.[71,72] Altseimer [47] and Bee Dagum and Heiss [73] (Mathematica) have used this data as a basis for correlating well cost as a function of depth, hole size, and rock type. A recent study used real cost information and drilling experience to estimate costs for sedimentary and hard crystalline rock to depths of 50,000 ft (15 km).[74] Cost data for drilling geothermal wells at The Geysers, the Imperial Valley, and at the dry hot rock demonstration site on the western rim of the Valles Caldera in New Mexico were also available.[47,48,75]

Our well cost model assumes that the functional relationship of cost with depth will be the same for geothermal wells as it is for oil and gas wells. Consequently, the existing geothermal well data can be extrapolated to cover a wide range of depth. Cost data were inflated to 1976 dollars to provide a consistent basis for comparison. Table C-1 shows well costs for the studies cited above [71,72,73,74] adjusted to 1976 dollars. Figure 17 was derived from the data plotted on Fig. C-1. A semilogarithmic correlation was used for convenience in extrapolation.

By using the Joint Assn. Surveys of 1972 [71] and 1973 [72] of the oil and gas industry, the Mathematica study [73] results for medium soft rock and the Continental Drilling estimates [74] as a basis for oil and gas wells, we developed the plot shown for oil and gas wells in Fig. C-1 or Fig. 17. The vapor-dominated plot was developed using The Geysers cost data as a reference point with the same dependence on depth assumed as was used for oil and gas wells. The same procedure was used again to establish the liquid-dominated and dry hot rock plot using Imperial Valley and LASL cost data as reference points. The Mathematica [73] predictions for hard and very hard rock seem to correlate well with this prediction. The Continental Drilling [74] estimates for hard rock do not follow the general trends shown in Fig. C-1 and are suspect mainly because of their weak dependence of cost on depth and because the costs do not include casing in the deep sections of the hole. Other cost information is available [56,77] but was not used in our model.

The geothermal well cost estimates for vapor-dominated, liquid-dominated, and dry hot rock systems will probably be conservatively high,particularly at depths greater than 4 km (13,000 ft). As the well gets deeper, geothermal drilling costs would be expected to merge closer to oil and gas well costs. However, until a large number of geothermal wells have been drilled at a variety of depths, we felt a conservative approach was justified.

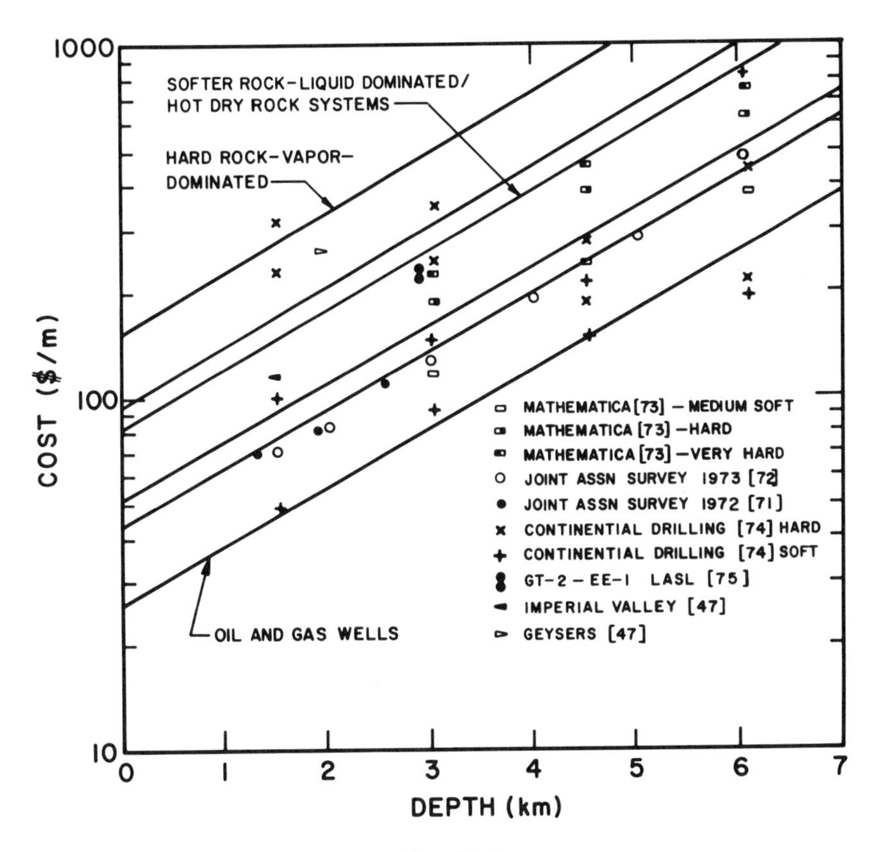

Fig. C-1
Well costs including drilling and casing as a function of depth. All cost data adjusted to 1976 dollars.

C.2 Reservoir Lifetime and Production Rates

The performance of a natural or artificially stimulated reservoir is perhaps the most difficult aspect of a geothermal system to predict. The well flow rates quoted in Section 6.2 represent ranges observed in the field. Given an accurate description of the reservoir including size, porosity, permeability, temperatures, and fluid volumes, the energy drawdown performance with or without reinjection could be predicted.[8] Unfortunately these variables are usually unknown or at least their changes with time uncertain. Field experiments, with the reservoir in question being tested, are really the only unequivocal record of performance. White and Williams and their associates at the U.S. Geological Survey [11] have developed models for predicting recoverable fractions of the available energy in a geothermal reservoir. Bloomster and associates [56] assume that the geothermal system lasts at least as long as the power plant itself, ~20 years. In some cases, field experience on

TABLE C-1

WELL COSTS ADJUSTED TO 1976 DOLLARS
(including drilling and casing)

Data Source	Depth (km)	Cost ($/m)	System	Rock Type
(1) Average values from 1972 Joint Association Survey[71,47]	1.3	70.00	oil and gas	Sedimentary
	1.9	83.80		
	2.6	118.50	6-9 5/8 in. dia	
(2) Average values from 1973 Joint Association Survey[72]	1.5	70.30	oil and gas	Sedimentary
	2.0	84.30		
	3.0	140.50		
	4.0	212.50	6-9 5/8 in. dia	
	5.0	300.70		
	6.0	444.90		
(3) Imperial Valley wells average values to depth[47]	1.6	102.90	liquid-dominated	Sedimentary-softer rock
(4) GT-2 and EE-1 holes Los Alamos Scientific Laboratory (LASL)[75]	2.9	230.00	dry hot rock geothermal 9 5/8 in. dia	Hard crystalline pre-Cambrian granite
(5) Geysers average values to depth[47]	1.8	270.00	vapor-dominated	Hard basaltic rock
(6) Mathematica[73]	3.05	101.78	oil and gas	soft - <1500 psi[a]
	4.57	211.28	6-12 in. ID	
3.05 km values are based on data, others are extrapolations	6.10	342.52		
	7.62	495.40		
"	3.05	115.42	"	medium soft-[a] 1500-3000 psi
	4.57	239.50		
	6.10	388.46		
	7.62	561.68		
"	3.05	131.76	"	medium hard-[a] 3000-8000 psi
	4.57	273.62		
	6.10	442.92		
	7.62	641.08		
"	3.05	183.92	"	hard-[a] 8000-16,000 psi
	4.57	381.88		
	6.10	618.76		
	7.62	895.02		
"	3.05	220.66	"	very hard- >16,000 psi[a]
	4.57	458.00		
	6.10	742.12		
	7.62	1073.50		
(7) Estimates from Continental Drilling[74]	1.52	46-93	8 1/2 - 9 5/8 in. dia	Sedimentary
	3.05	92-138		
	4.57	138-215		
	6.10	184-829		
	7.62	553-1014		
	9.14	769-1537		
"	1.52	230-323	8 1/2 - 9 5/8 in. dia	Hard crystalline rock
	3.05	219-346		
	4.57	185-277		
	6.10	207-461		
	7.62	553-830		
	9.14	615-1537		

[a] rock hardness levels from Hair[76]

producing a geothermal field has not occurred for more than a few years, and frequently with no decline in production rate or fluid quality over that period.[8] McNitt [78] and Towse [79] are optimistic about reservoir performance in California and particularly for liquid-dominated systems in the Imperial Valley area and would predict well lifetimes in excess of 20 years. Evidence at The Geysers [8] suggests a well lifetime of approximately 8 years to a 50% reduction in initial flow rate. The performance of vapor- and liquid-dominated systems is expected to differ.

An artificially stimulated, dry hot rock system might have some advantages over a natural system in that the reservoir characteristics and therefore performance may more easily be determined by early field performance tests. For example, McFarland [80] and Harlow and Pracht [81] have modeled fluid flow and heat transfer for a specified geometry crack. A thin, "penny-shaped" crack, circular in the crack plane and elliptical in cross-section is assumed. Heat is transferred by conduction through the rock to the circulating fluid contained within the crack, essentially as one-dimensional diffusive flow perpendicular to the crack plane. McFarland [80] has simplified fluid flow within the crack by assuming a two-dimensional circulation pattern with an energy source term at crack-fluid interface. In addition, he neglects thermal stress cracking and pressure impedance losses at the inlet and outlet regions. The coupled set of mass, momentum, and energy equations are solved in finite difference form. Figure *C-2* shows a solution for a 1500 m radius crack with rock and water properties listed in Table C-2. For a crack of this size, power extraction levels in excess of 60 MW (t) can be maintained indefinitely. Harlow and Pracht [81] illustrate the substantial enhancement of heat transfer rates when thermal stress cracking occurs. Thermal stress cracking could considerably reduce the fracture size required for a given power extraction rate.

C.3 Direct and Indirect Cost Factors for Wells

The direct cost factors for wells shown in Table 12 represent the additional costs for surface plumbing from the wellhead to the plant. Costs are given as a fraction of the well cost. Because estimates of actual well spacing are difficult to make, we assumed an equilateral triangular grid of wells spaced 200-300 m with the power plant centrally located. This technique, at least in a relative sense, provides a cost penalty for lower grade resources requiring more wells and surface plumbing.

The indirect cost factors of Table 13 are obviously important but very difficult to estimate. Our technique used Altseimer's [47] analysis of Greider's data [48] to extract indirect costs as a fraction of well costs. These fractions should be used for preliminary purposes only. For a given geothermal resource, the land acquisition, exploratory holes, and surface exploration costs could be expressed directly rather than as fractions of well cost.

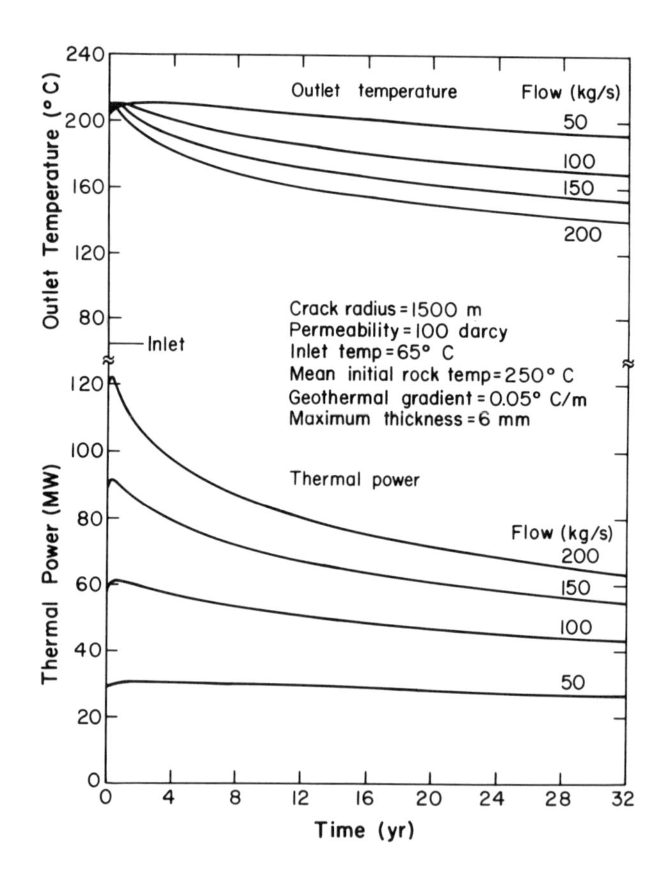

Fig. C-2
Power and temperature drawdown curves for a 1500 m radius crack with no thermal stress cracking (taken from Ref. 80).

TABLE C-2

ROCK AND WATER PROPERTIES

Rock[a]

ρ = 2700 kg/m^3 (168.6 lb/ft^3)

k = 2 W/mK (13.88 BTU/hr ft^2°F/in.)

C_p = 1000 J/kgK (0.24 BTU/lb°F)

$\beta = \frac{-1}{\rho}\left(\frac{\partial \rho}{\partial T}\right) = 8 \times 10^{-6} K^{-1}$

Water

ρ = 1000 kg/m^3 (62.4 lb/ft^3)

k = 0.63 W/mK (4.37 BTU/hr ft^2°F/in.)

C_p = 4200 J/kgK (1.0 BTU/lb°F)

$\mu = [1000 + 32(T-20)]^{-1}$ Pa-sec(T in °C)[b]

$\beta = \frac{-1}{\rho}\left(\frac{\partial \rho}{\partial T}\right) = 2 \times 10^{-4} K^{-1}$

---

[a] typical of granitic type rock
[b] 1 Pa-sec = 1000 centipoise (cp)

APPENDIX D

Economic Factors for Equipment

D.1 Heat exchanger and condenser costs

Heat exchanger and condenser costs are expressed as functions of tube and shell side pressures in Fig. 18. These cost curves resulted from the manufacturers' estimates shown in Fig. D-1. Because of the abnormally high rates of inflation for the past few years, it is difficult to extrapolate older cost data to present day costs. Therefore, we contacted manufacturers and obtained estimates for specific units. Furthermore, because unit sizes are from 20,000 to 30,000 ft^2 the economy of scale is not important and multiple units will be used with maximum tube length (>30 ft) employed to reduce costs. A reference design unit would have 3/4 to 1 in. diameter straight tubes, 30 to 40 ft long with a 1.25×diameter square or triangular pitch. One pass on shell and tube sides would be used with shell size and tube bundle size specified by shell side operating pressure for the shell and tube units, and by available air flow space for the air-cooled units.

In the primary heat exchanger, geothermal fluid would flow in the tubes which would have to be designed for chemical and mechanical removal of scale, that is with tube header box access. Working fluid would be heated on the shell side. If a water-cooled condenser was employed, working fluid would flow in the tubes and cooling water on the shell side. In some cases, aluminum-finned, air-cooled exchangers could be used for heat rejection with working fluid in the tubes.

Shell and tube side pressure effects on costs shown in Figs. D-1 and 18 are similar to those suggested by Peters and Timmerhaus [45] and Fraas and Ozisik.[53] A cost ratio between stainless steel and carbon steel, air-cooled units could be applied to shell and tube units at pressures below 600 psia. At higher pressures, fabrication costs are so high that the relative difference in materials cost decreases.

D.2 Pump costs

The correlation shown in Fig. 21 was derived from manufacturers' estimates of costs for multistage, centrifugal, high pressure, boiler type feed pumps. Cost data are plotted in Fig. D-2 using power output as the independent parameter as suggested by Peters and Timmerhaus [45] and Rudd and Watson.[46] The upper portion of the curve should be used for pressures above 2000 psi and the lower portion for pressures below 1000 psi.

D.3 Direct and indirect cost factors for equipment

The factored estimate method [46] employed requires fractional cost factors for various installational aspects of major equipment components. The indirect and direct cost fractions presented in Tables 12

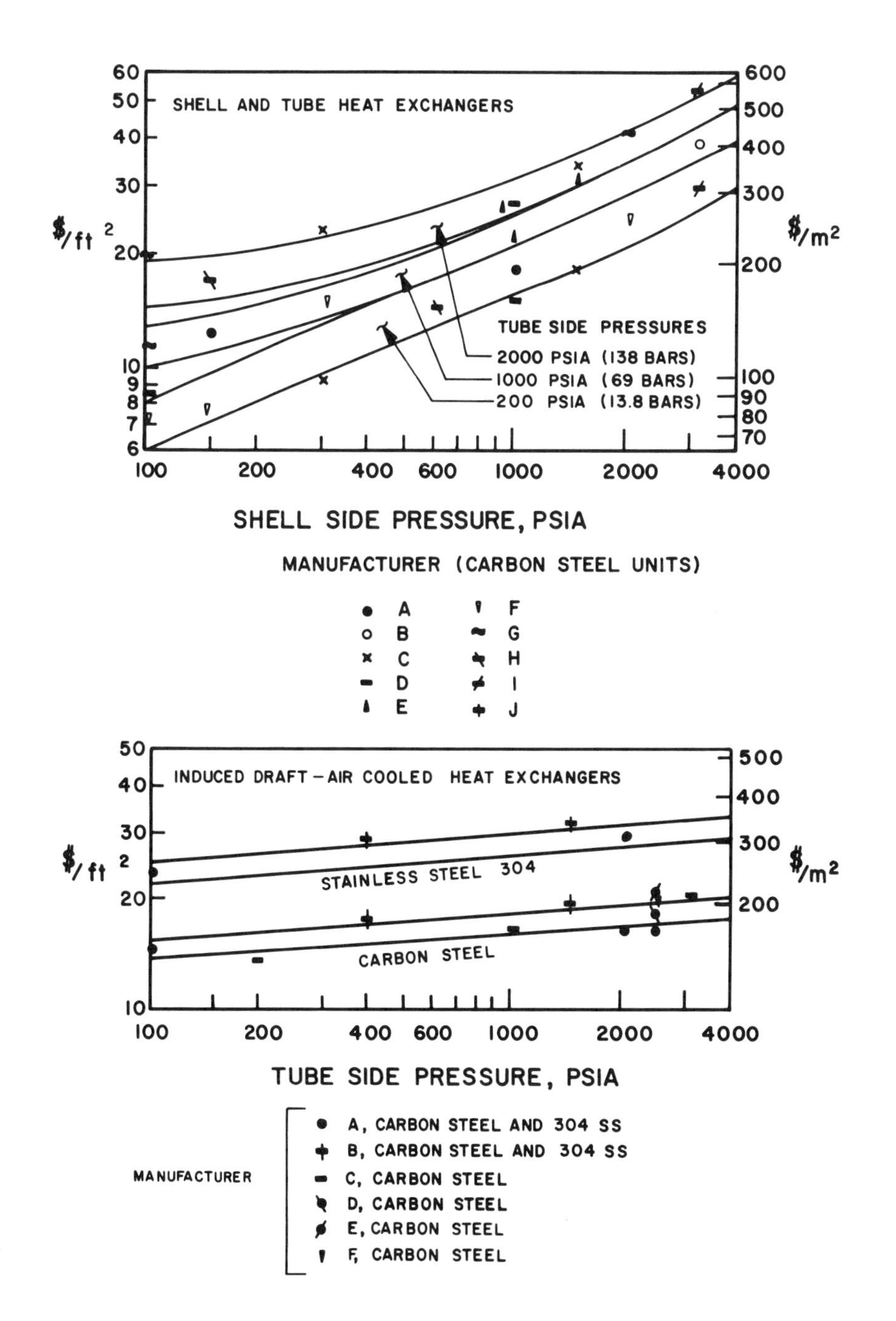

D-1

Heat exchanger and condenser costs — manufacturers' estimates of purchased cost in 1976 dollars. Unit size >20,000 ft^2. Air-cooled exchanger costs based on bare tube area.

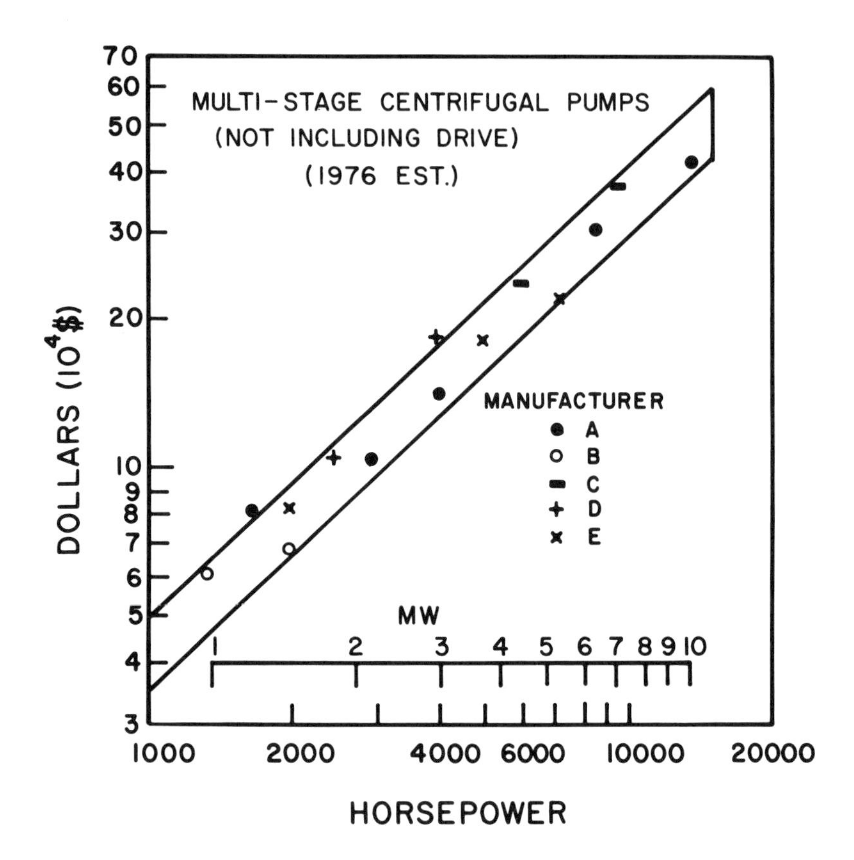

Fig. D-2
Pump costs — manufacturers' estimates of purchased cost in 1976 dollars.

and 13 were derived from those suggested by Rudd and Watson [46] and Peters and Timmerhaus [45] for chemical process plants by incorporating information presented by Fearnside and Cheney, [82] the Battelle report, [56] and Sesonske [64] for electric power plants. A 10% factor was also introduced for indirect costs associated with environmental impact assessment. Differences between fractions for vapor-, liquid-dominated, and dry hot rock systems are based on our estimates of distinct differences between these systems. Specific cases are described in Table 12.

APPENDIX E

Analytical Formulation of Thermodynamic Properties and Computerized Power Cycle Calculations

The Martin-Hou equation of state, Eq. (7), is expressed in a form that is convenient for obtaining analytical expressions for the various derived thermodynamic properties. This feature is desirable since it provides the opportunity for computer coding of the thermodynamic cycle analysis without the need to store large blocks of thermodynamic properties data.

When the expression for pressure is substituted into the integral relationships for enthalpy and entropy changes [Eqs. (12) and (13)] and the term-by-term integration is performed, the following algebraic equations result which relate the changes in enthalpy and entropy to the pressure, volume, and temperature of the thermodynamic state points between which the change takes place:

$$\frac{H_2(P_2,T_2) - H_1(P_1,T_1)}{RT_c} = \left\{C_o^* T_r + C_1^* \frac{T_r^2}{2} + C_2^* \frac{T_r^3}{3} + C_3^* \frac{T_r^4}{4}\right\}\Bigg|_{T_{r_1}}^{T_{r_2}}$$

$$+ \left\{Z_c P_r v_r - T_r + \sum_{i=1}^{4} \frac{f_i(1) - B_i + C_i\left[\left(1 + KT_r\right) e^{-KT_r} - e^{-K}\right]}{i\left(v_r - b\right)^i}\right.$$

$$+ \frac{f_5(1) - B_5 + C_5\left[\left(1 + KT_r\right) e^{-KT_r} - e^{-K}\right]}{a\, e^{av_r}}$$

$$\left. + \frac{f_6(1) - B_6 + C_6\left[\left(1 + KT_r\right) e^{-KT_r} - e^{-K}\right]}{2a\, e^{2av_r}}\right\}\Bigg|_{v_{r_1},T_{r_1}}^{v_{r_2},T_{r_2}} \qquad \text{(E-1)}$$

and

$$\frac{S_2 - S_1}{R} = \left\{ C_o^* \ell n\, T_r + C_1^* T_r + C_2^* \frac{T_r^2}{2} + C_3^* \frac{T_r^3}{3} - \ell n\, P_r \right\} \Bigg|_{P_{r_1}, T_{r_1}}^{P_{r_2}, T_{r_2}}$$

$$- \left\{ \ell n \frac{T_r}{(v_r - b) P_r Z_c} + \sum_{i=1}^{4} \frac{B_i - C_i K e^{-KT_r}}{i (v_r - b)^i} \right.$$

$$\left. + \frac{B_5 - C_5 K e^{-KT_r}}{a e^{a v_r}} + \frac{B_6 - C_6 K e^{-KT_r}}{2a\, e^{2 a v_r}} \right\} \Bigg|_{T_{r_1}, v_r(T_{r_1}, P_{r_1})}^{T_{r_2}, v_r(T_{r_2}, P_{r_2})}$$

(E-2)

The coefficients and constants that appear in these expressions correspond to those tabulated in Table 2 of Section 3.3. To evaluate enthalpy and entropy changes using Eqs. (E-1) and (E-2), appropriate values of temperature, pressure, and volume at state points 1 and 2 are required. Inasmuch as reduced volume appears implicitly in the Martin-Hou representation, it is necessary to solve iteratively for v_r whenever pressure and temperature are taken as primary thermodynamic variables. And, if any of the derived properties such as enthalpy or entropy are selected as primary variables (as is the case in heat exchange and turbine expansion calculations), then the determination of the remaining secondary variables such as temperature or pressure and volume is complicated by still another iterative calculation. While in principle the thermodynamic representation is analytical, in practice its use is limited by its complexity. Still, this approach is more convenient and, in many cases, more accurate than the alternative method of interpolation between individually stored data points (particularly for calculations in the vicinity of the critical point). Furthermore, the computer codes required to perform the thermodynamic analysis need not be overly complex. An example is the code from which the results of Sections 4.3 through 4.5 were obtained. This code, a listing of which appears at the end of this section, is partitioned for increased flexibility into two separate sections. The MAIN section manages all thermodynamic cycle calculations as described in Section 4.3, while the subroutine package performs, on command from MAIN, the thermodynamic properties evaluations using the various relationships and fluid properties introduced in Sections 3.2 and 3.3. Ammonia was selected for illustration in the FORTRAN listing; but other compounds can be substituted by simply inserting the appropriate constants from Table 2 into subroutines CSTNTS, LQDENS, PRSAT, and DPRSAT. Different geothermal/ambient conditions can be investigated by changing the appropriate variables in MAIN; namely the geothermal fluid temperature (TBRINE), the heat rejection temperature (TO), the turbine inlet temperature (TTOP), and the heat exchanger pinch-point temperature difference (DTPNCH). The turbine-inlet pressure is specified by assigning an integer value to the index N, which corresponds to the number of turbine stages, each with an assumed pressure ratio of 0.7. A single execution of the code provides sufficient information to complete a series of temperature-enthalpy diagrams like those shown in Fig. 4. In addition, well flow rates, utilization factors, thermodynamic cycle efficiencies, interstage turbine conditions, feed pump power, etc., are all included as output.

Other codes, more complicated yet similar in nature to the one described here, are now in the developmental state; and, when complete, these should be capable of providing estimates of both the thermodynamic and economic performance of geothermal power plants. For a description of such programs, the reader is referred to the recent work of Walter, [57] Bloomster et al., [56] and Green and Pines.[58] In addition, at least one code is available for obtaining Martin-Hou representations using the methods outlined in Section 3.2 [83]

TABLE E-1

Computer Listing for Power Cycle Calculation Code in Fortran IV

```
                                   (ENTRANCE)
                                       I
                                       I
*****************************************************************************
*      IMPLICIT REAL*8(A-H,O-Z)                                            *
*      DIMENSION FTC(6),B(6),C(6),CPO(4),TEMP(20),H(20),S(20),             *
*     1TR(20),SR(20),HR(20),T(20),TWATER(20)                               *
*      CALL CSTNTS(AM,TC,VC,PC,ZC,FTC,B,C,AR,BR,RK,CPO)                    *
*      TBRINE=150.D0                                                       *
*      TO=26.29D0                                                          *
*      TAMB=10.18D0                                                        *
*      EFFT=.85D0                                                          *
*      EFFP=.8D0                                                           *
*      TTOP=140.D0                                                         *
*      DTPNCH=10.D0                                                        *
*      PRMAX=2.D0                                                          *
*      RGAS=1.987D0*4184.D0/AM                                             *
*      VRC=1.D0                                                            *
*      TRC=1.D0                                                            *
*      PRC=1.D0                                                            *
*      N=7                                                                 *
*      TRTOP=(TTOP+273.15D0)*1.8D0/TC                                      *
*      CPWATR=4.2D3                                                        *
*C       GAS CONSTANT IS GIVEN IN JOULES PER KG-DEG. K                     *
*C       TEMPERATURES ARE IN DEG. C                                        *
*      RS=.7D0                                                             *
*      WRITE(6,1001)TBRINE,TAMB,TO,TTOP,DTPNCH,EFFT,EFFP,RS                *
*C       RS IS NOMINAL STAGE PRESSURE RATIO                                *
*      TRO=(TO+273.15D0)*1.8D0/TC                                          *
*      CALL PRSAT(TRO,PRO)                                                 *
*      CALL LQDENS(TRO,RHOO)                                               *
*      VRSLO=1.D0/RHOO                                                     *
*      CALL LATENT(TRO,HFGRO)                                              *
*      CALL VOLUME(VRSVO,TRO,PRO)                                          *
*      CALL HOFVT(HRSVO,VRSVO,TRO)                                         *
*      CALL SOFVT(SRSVO,VRSVO,TRO)                                         *
*      HRSLO=HRSVO-HFGRO                                                   *
*      SRSLO=SRSVO-HFGRO/TRO                                               *
*C       FIND THE MAX. NO. OF STAGES FOR SUBCRITICAL OPERATION.            *
*****************************************************************************
                                       I
                                       I
                                       I
*****************************************************************************
*      IF(PRO-RS.GT.0.)RS=PRO                                              *
*****************************************************************************
                                       I
                                       I
                                       I
*****************************************************************************
*      NMAX=IDINT(DLOG(PRO)/DLOG(RS))                                      *
*      PR1=PRO                                                             *
*****************************************************************************
                                       I
                                       I
```

```
      IF(TRTOP.LE.1.D0)N=NMAX

      DO 16 I=1,N

      NSTAGE=I
      PR1=PR1/RS
      P1=PR1*PC*1.03323D0
*C      PR1 IS REDUCED PRESSURE AFTER FEED PUMP.
      DELPR=PR1-PR0
      WRPUMP=ZC*DELPR*VRSL0/EFFP
      HRL1=HRSL0+WRPUMP
*C      HRL1 IS REDUCED ENTHALPY AFTER PUMP
      WRPIS=WRPUMP*EFFP
*C      WRPIS IS ISSENTROPIC PUMP WORK(REDUCED)
*C      REDUCED SPECIFIC PUMP WORK=(WORK/MASS)/(RGAS*TCRIT)
      DELHR=WRPUMP-WRPIS
      CALL DTRDH(DELHR,TR0,TRL1)
*C      TRL1 IS REDUCED TEMP. AFTER PUMP
      TR(1)=TRL1
      HR(1)=HRL1

      IF(I.GT.NMAX)GO TO 1

      CALL TRSAT(PR1,TRSL1)
*C      TRSL1 IS REDUCED SATURATED LIQUID TEMP. AT PR1

      IF(TRSL1.GT.TRTOP)GO TO 16

      TR(2)=TRSL1
      CALL VOLUME(VRSV1,TRSL1,PR1)
      CALL HOFVT(HRSV1,VRSV1,TRSL1)
*C      HRSV1 IS REDUCED SATURATED VAPOR ENTHALPY.
      CALL LATENT(TRSL1,HFGR1)
      HRSL1=HRSV1-HFGR1
*C      HRSL1 IS REDUCED SATURATED LIQUID ENTHALPY.
      CALL SOFVT(SRSV1,VRSV1,TRSL1)
      SRSL1=SRSV1-HFGR1/TRSL1
      HR(2)=HRSL1
      TR(3)=TR(2)
      HR(3)=HRSV1
      SR(3)=SRSV1

      GO TO 2
```

```
* 1     CONTINUE
*       CALL SOFVT(SRC,VRC,TRC)
*       CALL VOLUME(VRTOP,TRTOP,PR1)
*       CALL SOFVT(SRTOP,VRTOP,TRTOP)

*       IF(SRC.GT.SRTOP)GO TO 16

*       CALL HOFVT(HRC,VRC,TRC)
*       TR(2)=.975D0
*       TRD=TR(2)
*       CALL VOLUME(VRD,TRD,PR1)
*       CALL HOFVT(HRD,VRD,TRD)
*       HR(2)=HRD
*       CALL SOFVT(SRD,VRD,TRD)
*       SR(2)=SRD
*       CALL VOLUME(VR3,TRC,PR1)
*       CALL HOFVT(HR3,VR3,TRC)
*       CALL SOFVT(SR3,VR3,TRC)
*       HR(3)=HR3
*       SR(3)=SR3
*       TR(3)=TRC
```

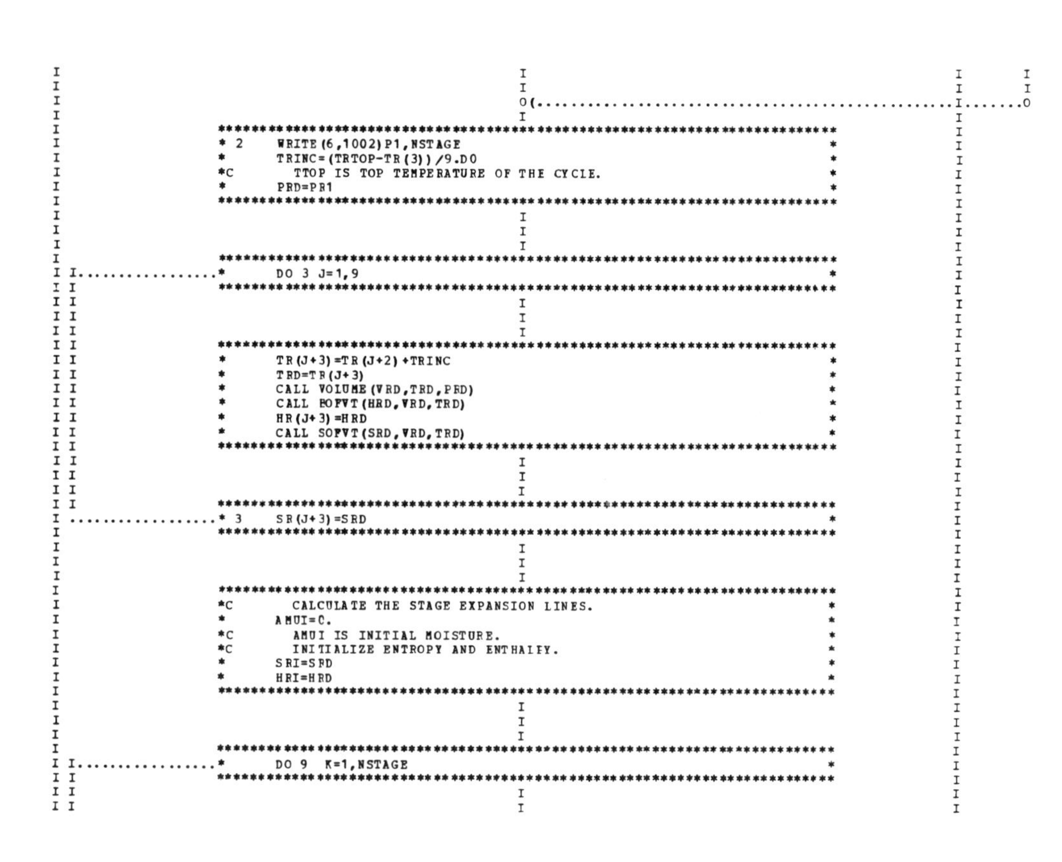

```
* 2     WRITE(6,1002)P1,NSTAGE
*       TRINC=(TRTOP-TR(3))/9.D0
*C        TTOP IS TOP TEMPERATURE OF THE CYCIE.
*       PRD=PR1

*       DO 3 J=1,9

*       TR(J+3)=TR(J+2)+TRINC
*       TRD=TR(J+3)
*       CALL VOLUME(VRD,TRD,PRD)
*       CALL HOFVT(HRD,VRD,TRD)
*       HR(J+3)=HRD
*       CALL SOFVT(SRD,VRD,TRD)

* 3     SR(J+3)=SRD

*C        CALCULATE THE STAGE EXPANSION LINES.
*       AMUI=C.
*C        AMUI IS INITIAL MOISTURE.
*C        INITIALIZE ENTROPY AND ENTHALPY.
*       SRI=SRD
*       HRI=HRD

*       DO 9  K=1,NSTAGE
```

```
                                         I                                           I
I I      *************************************************************************   I
I I      *     PRD=PRD*RS                                                        *   I
I I      *C      MOISTURE CHECK ON ISSENTROPIC EXPANSION TO THIS PRESSURE        *   I
I I      *     CALL TRSAT(PRD,TRSV)                                              *   I
I I      *     CALL VOIUME(VRSV,TRSV,PRD)                                        *   I
I I      *     CALL SOFVT(SRSV,VRSV,TRSV)                                        *   I
I I      *     CALL HOFVT(HRSV,VRSV,TRSV)                                        *   I
I I      *C      IF SR AT BEGINNING OF EXPANSION IS EQUAL OR GREATER             *   I
I I      *C      THAN SRSV, THE EXPANSION WILL BE DRY.                           *   I
I I      *************************************************************************   I
I I                                      I                                           I

I I                                      I                                           I
I I                                      I                                           I
I I                                      I                                           I
I I                                      I                                           I
I I      *************************************************************************   I
I I      *     IF(SRI-SRSV.GE.0.)GO TO 7                                         *......0   I
I I      *************************************************************************  I   I
I I                                      I                                          I   I
I I                                      I                                          I   I
I I                                      I                                          I   I
I I      *************************************************************************  I   I
I I      *     CALL LATENT(TRSV,HFGR)                                            *  I   I
I I      *     SRSL=SRSV-HFGR/TRSV                                               *  I   I
I I      *C      COMPUTE DEGREE OF MOISTURE FOR ISSENTROPIC EXPANSION            *  I   I
I I      *C      AND RESULTING ENTHALPY.                                         *  I   I
I I      *     AMUIS=(SRSV-SRI)*TRSV/HFGR                                        *  I   I
I I      *     HRIS=HRSV-AMUIS*HFGR                                              *  I   I
I I      *C      CALCULATE MOISTURE AFTER EXPANSION.                             *  I   I
I I      *     AMUEX=((HRI-HRIS)*(EFFT-AMUI/2.D0)+HRSV-HRI)/(HFGR+(HRI           *  I   I
I I      *    1-HRIS)/2.D0)                                                      *  I   I
I I      *************************************************************************  I   I
I I                                      I                                          I   I
I I                                      I                                          I   I
I I                                      I                                          I   I
I I      *************************************************************************  I   I
I I      *     IF(AMUEX.LE.0.)AMUEX=0.                                           *  I   I
I I      *************************************************************************  I   I
I I                                      I                                          I   I
I I                                      I                                          I   I
I I                                      I                                          I   I
I I      *************************************************************************  I   I
I I      *     AMUAV=(AMUI+AMUEX)/2.D0                                           *  I   I
I I      *     EFF=EFFT-AMUAV                                                    *  I   I
I I      *     HRDROP=EFF*(HRI-HRIS)                                             *  I   I
I I      *     HREX=HRI-HRDROP                                                   *  I   I
I I      *************************************************************************  I   I
I I                                      I                                          I   I
I I                                      I                                          I   I
I I                                      I                                          I   I
I I      *************************************************************************  I   I
I I      *  4  IF(AMUEX)5,5,6                                                    *......I...I.......0
I I      *************************************************************************  I   I       I
I I                                      I                                          I   I       I
I I                                      I                                          I   I       I
I I                                      I                                          I   I       I
I I      *************************************************************************  I   I       I
I I      *  5  HEX=HREX*1.987D0*TC/AM                                            *  I   I       I
I I      *     PD=PRD*PC*14.696D0                                                *  I   I       I
I I      *     CALL TEMPH(HEX,PD,TEX)                                            *  I   I       I
I I      *C********HEX,PD,AND TEX ARE IN ENGLISH UNITS                           *  I   I       I
I I      *     TREX=TEX/TC                                                       *  I   I       I
I I      *     CALL VOLUME(VREX,TREX,PRD)                                        *  I   I       I
I I      *     CALL SOFVT(SREX,VREX,TREX)                                        *  I   I       I
I I      *************************************************************************  I   I       I
I I                                      I                                          I   I       I
```

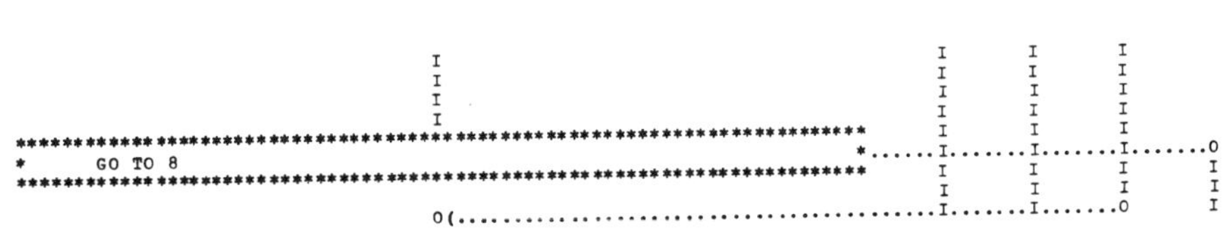

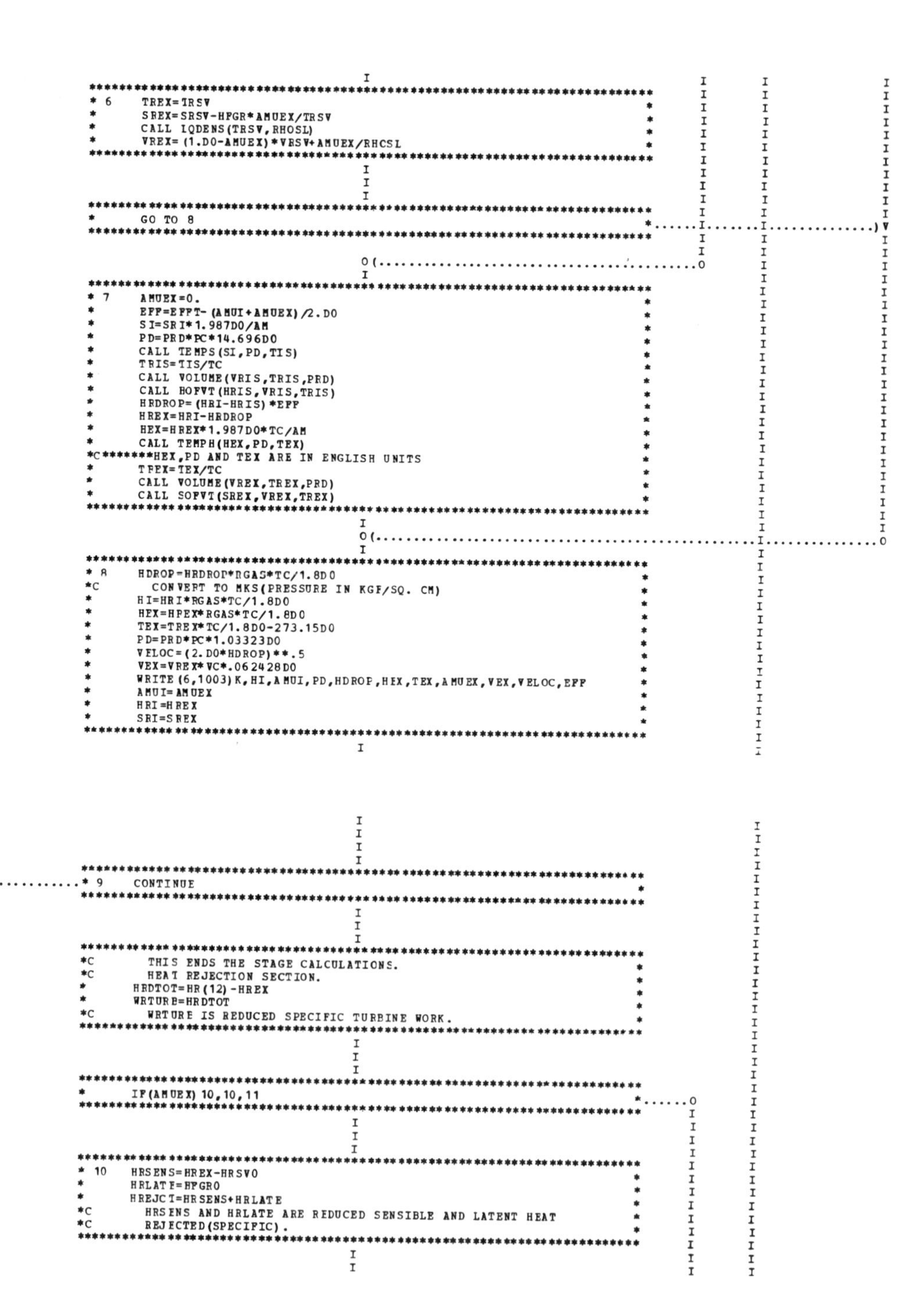

```
* 6    TREX=TRSV
*      SREX=SRSV-HFGR*AMUEX/TRSV
*      CALL LQDENS(TRSV,RHOSL)
*      VREX=(1.D0-AMUEX)*VRSV+AMUEX/RHCSL

*      GO TO 8

* 7    AMUEX=0.
*      EFF=EFFT-(AMUI+AMUEX)/2.D0
*      SI=SRI*1.987D0/AM
*      PD=PRD*PC*14.696D0
*      CALL TEMPS(SI,PD,TIS)
*      TRIS=TIS/TC
*      CALL VOLUME(VRIS,TRIS,PRD)
*      CALL HOFVT(HRIS,VRIS,TRIS)
*      HRDROP=(HRI-HRIS)*EFF
*      HREX=HRI-HRDROP
*      HEX=HREX*1.987D0*TC/AM
*      CALL TEMPH(HEX,PD,TEX)
*C*******HEX,PD AND TEX ARE IN ENGLISH UNITS
*      TREX=TEX/TC
*      CALL VOLUME(VREX,TREX,PRD)
*      CALL SOFVT(SREX,VREX,TREX)

* 8    HDROP=HRDROP*RGAS*TC/1.8D0
*C       CONVERT TO MKS(PRESSURE IN KGF/SQ. CM)
*      HI=HRI*RGAS*TC/1.8D0
*      HEX=HREX*RGAS*TC/1.8D0
*      TEX=TREX*TC/1.8D0-273.15D0
*      PD=PRD*PC*1.03323D0
*      VELOC=(2.D0*HDROP)**.5
*      VEX=VREX*VC*.062428D0
*      WRITE(6,1003)K,HI,AMUI,PD,HDROP,HEX,TEX,AMUEX,VEX,VELOC,EFF
*      AMUI=AMUEX
*      HRI=HREX
*      SRI=SREX

* 9    CONTINUE

*C       THIS ENDS THE STAGE CALCULATIONS.
*C       HEAT REJECTION SECTION.
*      HRDTOT=HR(12)-HREX
*      WRTURB=HRDTOT
*C       WRTURB IS REDUCED SPECIFIC TURBINE WORK.

*      IF(AMUEX) 10,10,11

* 10   HRSENS=HREX-HRSV0
*      HRLATE=HFGR0
*      HREJCT=HRSENS+HRLATE
*C       HRSENS AND HRLATE ARE REDUCED SENSIBLE AND LATENT HEAT
*C       REJECTED(SPECIFIC).
```

```
*       GO TO 12

* 11    HRSENS=0.
*C        THIS IS THE WET SECTION.
*       HRLATE=(1.D0-AMUEX)*HFGRO
*       HREJCT=HRSENS+HRLATE

* 12    CONTINUE
*       HRIN=HR(12)-HR(1)
*C        HRIN IS REDUCED HEAT INPUT(SPECIFIC).
*       WRNET=HRIN-HREJCT
*C        WRNET IS THE REDUCED SPECIFIC NET WORK OF CYCLE.
*       EFFCYC=WRNET/HRIN
*C        EFFCYC IS CYCLE EFFICIENCY.
*       PNET=1.D8
*C        PNET IS THE NET POWER NEEDED IN WATTS.
*       FLOW=1.8D0*PNET/(RGAS*TC*WRNET)
*C        FLOW IS THE NECESSARY COLD VAPOR FLOW TO OBTAIN PNET.
*       FACTOR=FLOW*RGAS*TC/1.8D0
*C        CALCULATE COMPONENT POWER AND HEAT.
*       PTURB=WRTURB*FACTOR
*       PPUMP=WRPUMP*FACTOR
*       HINEQ=HRIN*FACTOR
*       HSENEQ=HRSENS*FACTOR
*       HLATEQ=HRLATE*FACTOR
*C        CALCULATE THE HOT WATER PATH.
*       DHT=HRIN*RGAS*TC/1.8D0
*       HL=HR(1)*RGAS*TC/1.8D0
*       H(1)=HL
*       T(1)=TR(1)*TC/1.8D0-273.15D0
*       TL=TO+DTPNCH
*       TU=TBRINE

* 13    TAV=(TL+TU)/2.D0
*       SLOPE=(TBRINE-TAV)/DHT
*       DTMIN=TAV-(TR(1)*TC/1.8D0-273.15)
*       TWATER(1)=TAV

*       DO 14  L=2,12

*       H(L)=HR(L)*RGAS*TC/1.8D0
*       T(L)=TR(L)*TC/1.8D0-273.15
*       TWATER(L)=TWATER(1)+SLOPE*(H(L)-HL)
*       DELTAT=TWATER(L)-T(L)

* 14    IF(DELTAT.LT.DTMIN)DTMIN=DELTAT

*       IF(DABS(DTMIN-DTPNCH).LE..1D0)GO TO 15

*       IF((DTMIN-DTPNCH).LT.0.)TL=TAV
```

```
      IF((DTMIN-DTPNCH).GT.0.)TU=TAV

      GO TO 13

 15   CONTINUF
      WATFL=FLOW/(SLOPE*CPWATR)
C       WATFL IS TOTAL WATER FLOW RATE(KGM/SEC)
      YBRINE=WATFL*CPWATR*(TBRINE-TO-(TO+273.15D0)*DLOG((TBRINE
     1+273.15D0)/(TO+273.15D0)))
C       YBRINE IS TOTAL HOTWATER AVAIIABILITY IN MWE.
      YRES=WATFL*CPWATR*(TWATER(1)-TO-(TO+273.15D0)*DLOG((TWATER(1)
     1+273.15D0)/(TO+273.15D0)))
C       YRES IS THE RESIDUAL BRINE AVAIIABILITY.
      YHOT=YBRINE-YRES
      EFF1=PNET/YHOT
      EFF2=PNET/YBRINE
      WRITE(6,1004)PTURB,PPUMP,HINEQ,HSENEQ,HLATEQ,EFFCYC,FLOW,WATFL,
     1 YBRINE,YRES,EFF2
      WRITE(6,1005)TWATER(1),SLOPE
      WRITE(6,1006)(T(M),TWATER(M),H(M),M=1,12)

 16   CONTINUE

 1001 FORMAT(1H1,'BRINE TEMPERATURE(DEG.C)=',F10.3/1X,'AMBIENT',1X,
     1'TEMPERATURE=',F10.3/1X,'MINIMUM CYCLE TEMPERATURE=',
     2F10.3/1X,'TOP CYCLE TEMPERATURE=',F10.3/1X,'PINCH TEMF',
     3'ERATURE DIFFERENCE=',F10.3/1X,'DRY TURBINE EFFICIENCY=',
     4F10.3/1X,'FEED PUMP EFFICIENCY=',F10.3/1X,'NOMINAL',1X,
     5'STAGE PRESSURE RATIO=',F10.3///)
 1002 FORMAT(1H1,'MAXIMUM CYCLE PRESSURE(KGF/CM.SQ.)=',F10.3/1X,
     1'NUMBER OF STAGES=',I3//)
 1003 FORMAT(1X,'STAGE NO.',I3,/1X,'INITIAL ENTHALPY=',F15.3/
     11X,'INITIAL MOISTURE=',F10.3/1X,'EXIT PRESSURE(KGF/CM2)=',
     2F10.3/1X,'HEAT DROP=',F10.3/1X,'EXIT ENTHALPY=',F15.3/
     31X,'EXIT TEMPERATURE=',F10.3/1X,'EXIT MOISTURE=',F10.3/
     41X,'EXIT SPECIFIC VOLUME=',F10.3/1X,'NOZZLE VELOCITY=',
     5F10.3/1X,'STAGE EFFICIENCY=',F1C.3//)
 1004 FORMAT(1X,'NET TURBINE POWER=',E10.3/1X,'PUMPING POWER=',
     1E10.3/1X,'HEAT INPUT=',E10.3/1X,'SENSIBLE HEAT REJECTED=',
     2E10.3/1X,'LATENT HEAT REJECTED=',E10.3/1X,'NET CYCLE',1X,
     3'EFFICIENCY=',F10.3/1X,'WORKING FLUID FLOW=',E10.3//1X,
     4'WATER FLOW=',E10.3/1X,'BRINE AVAILABILITY=',E10.3/1X,
     5'RESIDUAL AVAILABILITY=',E10.3/1X,'CONVERSION EFFICIEN',
     6'CY(PNET/YBRINE)=',F10.3//)
 1005 FORMAT(1X,'RESIDUAL WATER TEMPERATURE=',F10.3/1X,'SLOPE',1X,
     1'OF WATER PATH(DEG.C PER J/KGM)=',E10.3///)
 1006 FORMAT(1X,'HEAT EXCHANGE PATH:'//1X,'FLUID TEMP',3X,
     1'WATER TEMP',4X,'ENTHALPY'//(1X,2F10.3,6X,F12.3))

      END
```

```
                                  (ENTRANCE)
                                      I
                                      I
******************************************************************************
*       SUBROUTINE CSTNTS(AM,TC,VC,PC,ZC,FTC,B,C,AR,BR,RK,CP)               *
*       IMPLICIT REAL*8(A-H,O-Z)                                             *
*       DIMENSION FTC(6),B(6),C(6),CP(4)                                     *
*C                                                                           *
*C        THERMODYNAMIC PROPERTIES OF AMMONIA                                *
*C                                                                           *
*C        THIS SUBROUTINE PROVIDES THE MARTIN-HOU EQUATION OF STATE          *
*C        COEFFICIENTS AS WELL AS THE IDEAL GAS SPECIFIC HEAT COEF-          *
*C        FICIENTS AND THE PROPERTIES OF THE COMPOUND AT THE THERMO-         *
*C        DYNAMIC CRITICAL POINT.                                            *
*C                                                                           *
*C        DEFINITION OF TERMS:                                               *
*C        TC = TEMPERATURE AT THE CRITICAL POINT IN DEGREES RANKINE.        *
*C        PC = PRESSURE AT THE CRITICAL POINT IN ATMOSPHERES.               *
*C        VC = SPECIFIC VOLUME AT THE CRITICAL POINT IN LBM./CUBIC FOOT.    *
*C        ZC = COMPRESSIBILITY FACTOR AT THE CRITICAL POINT.                *
*C        AM = MOLECULAR WEIGHT.                                             *
*C        CP(I) = REDUCED IDEAL GAS HEAT CAPACITY COEFFICIENTS.             *
*C                                                                           *
*C        THE DIMENSIONLESS MARTIN-HOU EQUATION OF STATE COEFFICIENTS ARE:*
*C        AR = VALUE OF REDUCED VOLUME EXPONENT.                             *
*C        BR = VALUE OF THE REDUCED 'EXCLUDED VOLUME TERM'.                  *
*C        RK = VALUE OF REDUCED TEMPERATURE EXPONENT.                        *
*C        FTC(I) = VALUE OF THE I'TH REDUCED TEMPERATURE FUNCTION EVAL-     *
*C        UATED AT THE CRITICAL TEMPERATURE.                                 *
*C        B(I) = VALUE OF REDUCED 'B' COEFFICIENT FOR THE I'TH REDUCED      *
*C        TEMPERATURE FUNCTION.                                              *
*C        C(I) = VALUE OF REDUCED 'C' COEFFICIENT  FOR THE I'TH REDUCED     *
*C        TEMPERATURE FUNCTION.                                              *
*C                                                                           *
*C        FOR ADDITIONAL INFORMATION SEE  S. L. MILORA, 'APPLICATION OF     *
*C        THE MARTIN EQUATION OF STATE TO THE THERMODYNAMIC PROPERTIES      *
*C        OF AMMONIA,' ORNL-TM-4413.                                         *
*C                                                                           *
*       TC=729.8D0                                                           *
*       VC=.06803D0                                                          *
*       ZC=.242D0                                                            *
*       PC=111.3D0                                                           *
*       CP(1)=4.30317D0                                                      *
*       CP(2)=-1.84109D0                                                     *
*       CP(3)=3.1995D0                                                       *
*       CP(4)=-.982953D0                                                     *
*       AR=14.33333D0                                                        *
*       BR=.01833333D0                                                       *
*       RK=6.4241900D0                                                       *
*       AM=17.032D0                                                          *
*       FTC(1)=-1.545953D0                                                   *
*       FTC(2)=1.022201D0                                                    *
*       FTC(3)=-.1891483D0                                                   *
*       FTC(4)=-4.589757D-2                                                  *
*       FTC(5)=1.741376D3                                                    *
*       FTC(6)=1.803382D5                                                    *
*       B(1)=.16586084D1                                                     *
*       B(2)=-.37680776D1                                                    *
*       B(3)=.49243365D1                                                     *
*       B(4)=-.19943707D1                                                    *
*       B(5)=.25646700D5                                                     *
*       B(6)=-.72299000D7                                                    *
*       C(1)=-.81254057D2                                                    *
*       C(2)=-.24544040D3                                                    *
*       C(3)=.64752053D3                                                     *
*       C(4)=-.32302139D3                                                    *
*       C(5)=.42946900D7                                                     *
*       C(6)=-.95713000D9                                                    *
******************************************************************************
                                      I
                                      I
                                      I
******************************************************************************
*       RETURN                                                               *
******************************************************************************


******************************************************************************
*       END                                                                  *
******************************************************************************
```

```
                                   (ENTRANCE)
                                        I
                                        I
**************************************************************************
*        SUBROUTINE VOLUME(VR,TR,PR)                                     *
*        IMPLICIT REAL*8(A-H,O-Z)                                        *
*        DIMENSION FTC(6),B(6),C(6),CP(4)                                *
*        DIMENSION F(6)                                                  *
*C                                                                       *
*C         SUBROUTINE VOLUME COMPUTES THE VALUE OF REDUCED VOLUME VR WHICH *
*C         CORRESPONDS TO THE GIVEN VALUES OF REDUCED TEMPERATURE TR AND *
*C         REDUCED PRESSURE PR. ALL QUANTITIES ARE THEREFORE DIMENSIONLESS.*
*C                                                                       *
*C         THE REDUCED TEMPERATURE IS DEFINED AS  TR = T/TC  WHERE T IS  *
*C         ABSOLUTE TEMPERATURE AND TC THE CRITICAL TEMPERATURE IN CON-  *
*C         SISTENT UNITS.                                                *
*C                                                                       *
*C         THE REDUCED PRESSURE IS DEFINED AS PR = P/PC WHERE P IS THE   *
*C         ABSOLUTE PRESSURE AND PC THE CRITICAL PRESSURE IN CONSISTENT  *
*C         UNITS.                                                        *
*C         THE REDUCED VOLUME IS DEFINED AS VR = V/VC WHERE V IS THE     *
*C         SPECIFIC VOLUME AND VC IS THE CRITICAL VALUE IN CONSISTENT    *
*C         UNITS.                                                        *
*C                                                                       *
*C         VOLUME CALLS SUBROUTINE CSTNTS TO OBTAIN THE MARTIN-HOU       *
*C         EQUATION OF STATE COEFFICIENTS FOR THE COMPOUND.              *
*C                                                                       *
*C         IF THE REDUCED TEMPERATURE IS LESS THAN 1.0, VOLUME CALLS     *
*C         SUBROUTINE PRSAT TO CHECK WHETHER THE GIVEN VALUES OF TR      *
*C         AND VR  CORRESPOND TO SUPERHEATED VAPOR OR SUBCOOLED LIQUID.  *
*C                                                                       *
*C         SINCE THE MARTIN-HOU EQUATION IS NOT EXPLICIT IN VR, AN ITERA- *
*C         TIVE  SOLUTION IS REQUIRED. SUCCESSIVE APPROXIMATIONS TO VR ARE *
*C         CALCULATED ON THE ISOTHERM TR UNTIL THE CALCULATED VALUE OF   *
*C         REDUCED PRESSURE, PRCAL, APPROACHES THE GIVEN VALUE, PR, TO   *
*C         WITHIN A SPECIFIED TOLERENCE EPS.                             *
*C                                                                       *
*        CALL CSTNTS(AM,TC,VC,PC,ZC,FTC,B,C,AR,BR,RK,CP)                 *
**************************************************************************
                                        I
                                        I
                                        I
**************************************************************************
I..................*        DO 1 M=1,6                                   *
I                 ********************************************************
I                                       I
I                                       I
I                                       I
I                 ********************************************************
...................* 1      F(M)=FTC(M)+B(M)*(TR-1.DO)+C(M)*(DEXP(-TR*RK)-DEXP(-RK))  *
                  ********************************************************
                                        I
                                        I
                                        I
**************************************************************************
*        EPS=1.D-6                                                       *
**************************************************************************
                                        I



                                        I
                                        I
                                        I
                                        I
**************************************************************************
*        IF(TR-1.DO)2,7,7                                                *......O
**************************************************************************      I
                                        I                                       I
                                        I                                       I
                                        I                                       I
**************************************************************************      I
* 2      CALL PRSAT(TR,PRS)                                              *      I
**************************************************************************      I
                                        I                                       I
                                        I                                       I
                                        I                                       I
**************************************************************************      I
*        IF((PR-PRS).GE.1.D-6) GO TO 5                                   *......I......O
**************************************************************************      I      I
                                        I                                       I      I
                                        I                                       I      I
                                        I                                       I      I
**************************************************************************      I      I
*        IF(DABS(PR-PRS).LE.1.D-6) EPS=3.D-4                             *      I      I
**************************************************************************      I      I
                                        I                                       I      I
```

```
*       VU=TR/(ZC*PR)
*       VL=ZC*VU
*       TERM=1.D0/(VL-BR)
*       VR=VL
*       DPDV=(-1.D0/ZC)*TERM*TERM*(TR+TERM*(2.D0*F(1)+TERM*(3.D0*F(2)+
*      1TERM*(4.D0*F(3)+TERM*5.D0*F(4)))))-(AR*DEXP(-AR*VR)/ZC)*(F(5)+
*      22.D0*F(6)*DEXP(-AR*VR))

*       IF(DPDV.LE.0.)GO TO 13

*       VMEAN=(VU+VL)/2.D0

* 3     VR=VMEAN
*       TERM=1.D0/(VMEAN-BR)
*       PRCAL=(1.D0/ZC)*TERM*(TR+TERM*(F(1)+TERM*(F(2)+TERM
*      1*(F(3)+TERM*(F(4))))))+(1.D0/ZC)*DEXP(-AR*VR)*(F(5)+
*      2F(6)*DEXP(-AR*VR))
*       DPDV=(-1.D0/ZC)*TERM*TERM*(TR+TERM*(2.D0*F(1)+TERM*(3.D0*F(2)+
*      1TERM*(4.D0*F(3)+TERM*5.D0*F(4)))))-(AR*DEXP(-AR*VR)/ZC)*(F(5)+
*      22.D0*F(6)*DEXP(-AR*VR))
*       DEL=(PR-PRCAL)/PR

*       IF(DABS(DEL).LE.EPS) GO TO 17

*       IF(DABS(PR-PRS).GE.1.D-5) GO TO 4

*       IF(DABS(DPDV).LE.1.D-5)GO TO 17

* 4     CONTINUE

*       IF(DEL.LE.0.OR.DPDV.GE.0.)VL=VMEAN

*       IF(DEL.GE.0.AND.DPDV.LE.0.)VU=VMEAN

*       VMEAN=(VU+VL)/2.D0

*       GO TO 3
```

```
 5      VU=1.D0
        VL=.3D0
        TERM=1.DO/(VU-BR)
        VR=VU
        DPDV=(-1.DO/ZC)*TERM*TERM*(TR+TERM*(2.DO*F(1)+TERM*(3.DO*F(2)+
       1TERM*(4.DO*F(3)+TERM*5.DO*F(4)))))-(AR*DEXP(-AR*VR)/ZC)*(F(5)+
       22.DO*F(6)*DEXP(-AR*VR))

        IF(DPDV.LE.0.)GO TO 14

        VMEAN=(VU+VL)/2.D0

 6      VR=VMEAN
        TERM=1.DO/(VMEAN-BR)
        PRCAL=(1.DO/ZC)*TERM*(TR+TERM*(F(1)+TERM*(F(2)+TERM
       1*(F(3)+TERM*(F(4))))))+(1.DO/ZC)*DEXP(-AR*VR)*(F(5)+
       2F(6)*DEXP(-AR*VR))
        DPDV=(-1.DO/ZC)*TERM*TERM*(TR+TERM*(2.DO*F(1)+TERM*(3.DO*F(2)+
       1TERM*(4.DO*F(3)+TERM*5.DO*F(4)))))-(AR*DEXP(-AR*VR)/ZC)*(F(5)+
       22.DO*F(6)*DEXP(-AR*VR))
        DEL=(PR-PRCAL)/PR

        IF(DABS(DEL).LE.EPS) GO TC 17

        IF(DEL.GE.0.OR.DPDV.GE.0.)VU=VMEAN

        IF(DEL.LE.0.AND.DPDV.LE.0.)VL=VMEAN

        VMEAN=(VU+VL)/2.D0

        GO TO 6
```

```
                                                      I  I  I     I
                                                      I  I  I     I
                         O(...........................O  I  I     I
                         I                                I  I     I
**************************************************       I  I     I
* 7   VL=.3D0                                    *       I  I     I
*     VU=TR/(ZC*PR)                              *       I  I     I
**************************************************       I  I     I
                         I                               I  I     I
                         O(...........................O  I  I     I
                         I                            I  I  I     I
**************************************************    I  I  I     I
* 8   CALL PRES(PRU,TR,VU)                       *    I  I  I     I
**************************************************    I  I  I     I
                         I                            I  I  I     I
                         I                            I  I  I     I
                         I                            I  I  I     I
**************************************************    I  I  I     I
*     IF(PRU-PR)10,9,9                           *....I..I..I.....I.....O
**************************************************    I  I  I     I     I
                         I                            I  I  I     I     I
                         I                            I  I  I     I     I
                         I                            I  I  I     I     I
**************************************************    I  I  I     I     I
* 9   VU=1.5*VU                                  *    I  I  I     I     I
**************************************************    I  I  I     I     I
                         I                            I  I  I     I     I
                         I                            I  I  I     I     I
                         I                            I  I  I     I     I
**************************************************    I  I  I     I     I
*     GO TO 8                                    *....O  I  I     I     I
**************************************************       I  I     I     I
                                                         I  I     I     I
                         O(..............................I..I.....I.....O
                         I                               I  I     I     I
**************************************************       I  I     I     I
* 10  CALL PRES(PRL,TR,VL)                       *       I  I     I     I
**************************************************       I  I     I     I
                         I                               I  I     I     I
                         I                               I  I     I     I
                         I                               I  I     I     I
**************************************************       I  I     I     I
*     IF(PRL-PR)11,11,12                         *....O  I  I     I     I
**************************************************    I  I  I     I     I
                         I                            I  I  I     I     I
                         I                            I  I  I     I     I
                         I                            I  I  I     I     I
**************************************************    I  I  I     I     I
* 11  VL=VL/1.5D0                                *    I  I  I     I     I
**************************************************    I  I  I     I     I
                         I                            I  I  I     I     I
                         I                            I  I  I     I     I
                         I                            I  I  I     I     I
**************************************************    I  I  I     I     I
*     GO TO 10                                   *....I..I..I.....I.....O
**************************************************    I  I  I     I
                                                      I  I  I     I
```

```
                                                      I  I  I     I
                                                      I  I  I     I
                         O(...........................O  I  I     I
                         I                               I  I     I
**************************************************       I  I     I
* 12  CONTINUE                                   *       I  I     I
**************************************************       I  I     I
                         I                               I  I     I
                         I                               I  I     I
                         I                               I  I     I
**************************************************       I  I     I
*     GO TO 15                                   *....O  I  I     I
**************************************************    I  I  I     I
                                                      I  I  I     I
                         O(...........................I..I..I.....O
                         I                            I  I  I
**************************************************    I  I  I
* 13  VU=TR/(ZC*PR)                              *    I  I  I
*     VL=ZC*VU                                   *    I  I  I
**************************************************    I  I  I
                         I                            I  I  I
                         I                            I  I  I
                         I                            I  I  I
**************************************************    I  I  I
*     GO TO 15                                   *...)V  I  I
**************************************************    I  I  I
                                                      I  I  I
```

```
                                          O(..............................I...I...O
                                          I                               I   I
**********************************************************************   I   I
* 14   VU=1.D0                                                       *   I   I
*      VL=.3D0                                                       *   I   I
**********************************************************************   I   I
                                          I                               I   I
                                          I                               I   I
                                          I                               I   I
**********************************************************************   I   I
*      GO TO 15                                                      *   I   I
**********************************************************************   I   I
                                          I                               I   I
                                          O(..............................O   I
                                          I                                   I
**********************************************************************       I
* 15   VMEAN=(VU+VL)/2.D0                                            *       I
**********************************************************************       I
                                          I                                   I
                                          O(..............................O   I
                                          I                               I   I
**********************************************************************   I   I
* 16   VR=VMEAN                                                      *   I   I
*      TERM=1.D0/(VR-BR)                                             *   I   I
*      PRCAL=(1.D0/ZC)*TERM*(TR+TERM*(F(1)+TERM*(F(2)+TERM           *   I   I
*     1*(F(3)+TERM*(F(4))))))+(1.D0/ZC)*DEXP(-AR*VR)*(F(5)+          *   I   I
*     2F(6)*DEXP(-AR*VR))                                            *   I   I
*      DEL=(PR-PRCAL)/PR                                             *   I   I
**********************************************************************   I   I
                                          I                               I   I


                                          I                               I   I
                                          I                               I   I
                                          I                               I   I
                                          I                               I   I
**********************************************************************   I   I
*      IF(DABS(DEL).LE.EPS) GO TO 17                                 *......I..)V
**********************************************************************   I   I
                                          I                               I   I
                                          I                               I   I
                                          I                               I   I
**********************************************************************   I   I
*      IF(DEL.GE.0.) VU=VMEAN                                        *   I   I
**********************************************************************   I   I
                                          I                               I   I
                                          I                               I   I
                                          I                               I   I
**********************************************************************   I   I
*      IF(DEL.LE.0.) VL=VMEAN                                        *   I   I
**********************************************************************   I   I
                                          I                               I   I
                                          I                               I   I
                                          I                               I   I
**********************************************************************   I   I
*      VMEAN=(VL+VU)/2.D0                                            *   I   I
**********************************************************************   I   I
                                          I                               I   I
                                          I                               I   I
                                          I                               I   I
**********************************************************************   I   I
*      GO TO 16                                                      *......O   I
**********************************************************************       I
                                          I                                   I
                                          O(..................................O
                                          I
**********************************************************************
* 17   VR=VMEAN                                                      *
**********************************************************************
                                          I
                                          I
                                          I
**********************************************************************
*      RETURN                                                        *
**********************************************************************


**********************************************************************
*      END                                                           *
**********************************************************************
```

```
                              (ENTRANCE)
                                  I
                                  I
      **********************************************************************
      *      SUBROUTINE HDEV(DELTH,TR,PR,VR)                                *
      *      IMPLICIT REAL*8(A-H,O-Z)                                       *
      *      DIMENSION FTC(6),B(6),C(6),CP(4)                               *
      *      DIMENSION VIRIAL(6),FINV(6)                                    *
      *C                                                                    *
      *C       GIVEN THE REDUCED TEMPERATURE, PRESSURE, AND VOLUME (TR,PR,VR), *
      *C       HDEV CALCULATES THE DEPARTURE OF THE REDUCED ENTHALPY FROM   *
      *C       THE CORRESPONDING IDEAL GAS VALUE.                           *
      *C                                                                    *
      *C       DEFINITION OF TERMS:                                         *
      *C       DELTH = REDUCED DEPARTURE OF ENTHALPY FROM IDEAL GAS VALUE.  *
      *C       TR = REDUCED TEMPERATURE.= T/TC.                             *
      *C       PR = REDUCED PRESSURE = P/PC.                                *
      *C       VR = REDUCED VOLUME = V/VC.                                  *
      *C                                                                    *
      *C       ALL QUANTITIES ARE DIMENSIONLESS.                            *
      *C                                                                    *
      *C       THE ENTHALPY DEPARTURE IN ANY UNITS WOULD BE GIVEN BY THE    *
      *C       PRODUCT DELTH*CRITICAL TEMPERATURE *GAS CONSTANT.            *
      *C                                                                    *
      *C       HDEV CALLS SUBROUTINE CSTNTS TO OBTAIN THE REDUCED MARTIN-HOU *
      *C       EQUATION OF STATE COEFFICIENTS.                              *
      *C                                                                    *
      *C       TO OBTAIN THE REDUCED VALUE OF ENTHALPY FOR THE REAL GAS, DELTH *
      *C       MUST BE SUBTRACTED FROM THE IDEAL GAS CONTRIBUTION.          *
      *C                                                                    *
      *      CALL CSTNTS(AM,TC,VC,PC,ZC,FTC,B,C,AR,BR,RK,CP)                *
      *      TERM=1.DO/(VR-BR)                                              *
      *      E=DEXP(-RK)                                                    *
      *      ET=DEXP(-RK*TR)                                                *
      *      DELTH=TR-ZC*PR*VR                                              *
      **********************************************************************
                                  I
                                  I
                                  I
      **********************************************************************
I.....*      DO1 I=1,4                                                      *
I     **********************************************************************
I                                 I
I                                 I
I                                 I
I     **********************************************************************
I     *      VIRIAL(I)=FTC(I)-B(I)+C(I)*(ET*(1.DO+RK*TR)-E)                 *
I     *      FINV(I)=TERM**I/I                                              *
I     **********************************************************************
I                                 I
I                                 I
I                                 I
I     **********************************************************************
......* 1    DELTH=DELTH-VIRIAL(I)*FINV(I)                                  *
      **********************************************************************
                                  I
                                  I
                                  I
                                  I
                                  I
      **********************************************************************
I.....*      DO 2 J=1,2                                                     *
I     **********************************************************************
I                                 I
I                                 I
I                                 I
I     **********************************************************************
I     *      VIRIAL(J+4)=FTC(J+4)-B(J+4)+C(J+4)*(ET*(1.DO+RK*TR)-E)         *
I     *      FINV(J+4)=DEXP(-J*AR*VR)/(J*AR)                                *
I     **********************************************************************
I                                 I
I                                 I
I                                 I
I     **********************************************************************
......* 2    DELTH=DELTH-VIRIAL(J+4)*FINV(J+4)                              *
      **********************************************************************
                                  I
                                  I
                                  I
      **********************************************************************
      *      RETURN                                                         *
      **********************************************************************


      **********************************************************************
      *      END                                                            *
      **********************************************************************
```

```
                                        (ENTRANCE)
                                            I
                                            I
**************************************************************************
*       SUBROUTINE ENTHIG(TR,HIG)                                        *
*       IMPLICIT REAL*8(A-H,O-Z)                                         *
*       DIMENSION FTC(6),B(6),C(6),CP(4)                                 *
*C                                                                       *
*C        ENTHIG CALCULATES THE VALUE OF REDUCED ENTHALPY FOR THE IDEAL  *
*C        GAS AT REDUCED TEMPERATURE TR BY INTEGRATING THE THIRD ORDER   *
*C        POWER SERIES EXPRESSION FOR THE REDUCED IDEAL GAS SPECIFIC     *
*C        HEAT FUNCTION.  THE REFFERENCE IS ARBITRARILY CHOSEN TO BE ZERO *
*C        AT TR=0.                                                       *
*C                                                                       *
*C        ALL QUANTITIES ARE DIMENSIONLESS BY VIRTURE OF THEIR BEING     *
*C        REDUCED.                                                       *
*C                                                                       *
*C        ENTHIG CALLS  SUBROUTINE CSTNTS TO OBTAIN THE REDUCED IDEAL GAS *
*C        HEAT CAPACITY COEFFICIENTS.                                    *
*C                                                                       *
*C        DEFINITION OF TERMS:                                           *
*C        TR= T/TC = REDUCED TEMPERATURE.                                *
*C        HIG = REDUCED IDEAL GAS ENTHALPY.                              *
*C                                                                       *
*C        THE IDEAL GAS ENTHALPY IN ANY UNITS IS GIVEN BY THE PRODUCT    *
*C        HIG*CRITICAL TEMPERATURE*GAS CONSTANT.                         *
*C                                                                       *
*       CALL CSTNTS(AM,TC,VC,PC,ZC,FTC,B,C,AR,BR,RK,CP)                  *
*       HIG=CP(1)*TR                                                     *
**************************************************************************
                                            I
                                            I
                                            I
**************************************************************************
I..................*       DO 1 I=2,4                                                       *
I                 **************************************************************************
I                                                           I
I                                                           I
I                                                           I
I                 **************************************************************************
...................* 1     HIG=HIG+CP(I)*TR**I/I                                            *
                  **************************************************************************
                                            I
                                            I
                                            I
**************************************************************************
*       RETURN                                                           *
**************************************************************************


**************************************************************************
*       END                                                              *
**************************************************************************
```

```
                                        (ENTRANCE)
                                            I
                                            I
**************************************************************************
*       SUBROUTINE ENTRIG(TR,PR,SIG)                                     *
*       IMPLICIT REAL*8(A-H,O-Z)                                         *
*       DIMENSION FTC(6),B(6),C(6),CP(4)                                 *
*C                                                                       *
*C        ENTRIG EVALUATES THE EXPRESSION FOR THE IDEAL GAS ENTROPY GIVEN *
*C        REDUCED TEMPERATURE TR AND THE REDUCED PRESSURE PR. LIKE       *
*C        ENTRIG, IT REQUIRES THE REDUCED IDEAL GAS HEAT CAPACITY        *
*C        COEFFICIENTS WHICH ARE SUPPLIED BY  SUBROUTINE CSTNTS.         *
*C                                                                       *
*C        ALL QUANTITIES ARE DIMENSIONLESS.                              *
*C                                                                       *
*C        DEFINITION OF TERMS:                                           *
*C        SIG = REDUCED IDEAL GAS ENTROPY.                               *
*C        TR =  REDUCED TEMPERATURE.                                     *
*C        PR =  REDUCED PRESSURE.                                        *
*C                                                                       *
*C        SINCE NO REFFERENCE STATE IS REQUIRED, THE DIFFERENCE IN IDEAL *
*C        GAS ENTROPY BETWEEN TWO STATE POINTS IS FOUND BY EVALUATING    *
*C        ENTRIG AT THE TWO POINTS AND SUBTRACTING THE RESULTS.          *
*C        THE IDEAL GAS ENTROPY IN ANY UNITS IS GIVEN BY THE  PRODUCT    *
*C        OF SIG AND THE GAS CONSTANT.                                   *
*C                                                                       *
*       CALL CSTNTS(AM,TC,VC,PC,ZC,FTC,B,C,AR,BR,RK,CP)                  *
*       SIG=CP(1)*DLOG(TR)-DLOG(PR)                                      *
**************************************************************************
                                            I
                                            I
```

```
                                                                  I
                          ***************************************************************************
I.........................*      DO 1 I=2,4                                                         *
I                         ***************************************************************************
I                                                                 I
I                                                                 I
I                                                                 I
I                         ***************************************************************************
..........................* 1    SIG=SIG+CP(I)*TR**(I-1)/(I-1)                                      *
                          ***************************************************************************
                                                                  I
                                                                  I
                                                                  I
                          ***************************************************************************
                          *      RETURN                                                             *
                          ***************************************************************************


                          ***************************************************************************
                          *      END                                                                *
                          ***************************************************************************
```

```
                                                              (ENTRANCE)
                                                                  I
                                                                  I
                          ***************************************************************************
                          *      SUBROUTINE SDEV(DELTS,TR,PR,VR)                                    *
                          *      IMPLICIT REAL*8(A-H,O-Z)                                           *
                          *      DIMENSION FTC(6),B(6),C(6),CP(4)                                   *
                          *      DIMENSION VIRIAL(6),FINV(6)                                        *
                          *C                                                                        *
                          *C       GIVEN THE REDUCED TEMPERATURE, PRESSURE AND VOLUME (TR,PR,VR),   *
                          *C       SDEV CALCULATES THE DEPARTURE OF THE REDUCED ENTROPY FROM THE    *
                          *C       CORRESPONDING IDEAL GAS VALUE IN A MANNER ANALAGOUS TO HDEV.     *
                          *C                                                                        *
                          *C       DEFINITION OF TERMS:                                             *
                          *C       DELTS = REDUCED VALUE OF ENTROPY DEPARTURE FROM IDEAL GAS VALUE.*
                          *C       TR = T/TC = REDUCED TEMPERATURE.                                 *
                          *C       PR = P/PC = REDUCED PRESSURE.                                    *
                          *C       VR = V/VC = REDUCED VOLUME.                                      *
                          *C                                                                        *
                          *C       ALL QUANTITIES ARE DIMENSIONLESS.                                *
                          *C                                                                        *
                          *C       THE ENTROPY DEPARTURE IN ANY UNITS WOULD BE GIVEN BY THE         *
                          *C       PRODUCT DELTS*GAS CONSTANT.                                      *
                          *C                                                                        *
                          *C       SDEV CALLS SUBROUTINE CSTNTS TO OBTAIN THE REDUCED MARTIN-HOU    *
                          *C       EQUATION OF STATE COEFFICIENTS.                                  *
                          *C                                                                        *
                          *C       TO OBTAIN THE REDUCED VALUE OF  ENTROPY FOR THE REAL GAS, DELTS  *
                          *C       MUST BE SUBTRACTED FROM THE CONTRIBUTION FROM THE IDEAL GAS.     *
                          *C                                                                        *
                          *      CALL CSTNTS(AM,TC,VC,PC,ZC,FTC,B,C,AR,BR,RK,CP)                    *
                          *      TERM=1.D0/(VR-BR)                                                  *
                          *      E=DEXP(-RK)                                                        *
                          *      ET=DEXP(-RK*TR)                                                    *
                          *      DELTH=TR-ZC*PR*VR                                                  *
                          ***************************************************************************
                                                                  I
                                                                  I
                                                                  I
                          ***************************************************************************
I.........................*      DO1 I=1,4                                                          *
I                         ***************************************************************************
I                                                                 I
I                                                                 I
I                                                                 I
I                         ***************************************************************************
I                         *      VIRIAL(I)=FTC(I)-B(I)+C(I)*(ET*(1.D0+RK*TR)-E)                     *
I                         *      FINV(I)=TERM**I/I                                                  *
I                         ***************************************************************************
I                                                                 I
I                                                                 I
I                                                                 I
I                         ***************************************************************************
..........................* 1    DELTH=DELTH-VIRIAL(I)*FINV(I)                                      *
                          ***************************************************************************
                                                                  I
```

```
                                      I
                                      I
                                      I
                                      I
          ***********************************************************
I.........*      DO 2 J=1,2                                         *
I         ***********************************************************
I                                     I
I                                     I
I                                     I
I         ***********************************************************
I         *      VIRIAL(J+4)=FTC(J+4)-B(J+4)+C(J+4)*(ET*(1.D0+RK*TR)-E) *
I         *      FINV(J+4)=DEXP(-J*AR*VR)/(J*AR)                     *
I         ***********************************************************
I                                     I
I                                     I
I                                     I
I         ***********************************************************
..........* 2    DELTH=DELTH-VIRIAL(J+4)*FINV(J+4)                  *
          ***********************************************************
                                      I
                                      I
                                      I
          ***********************************************************
          *      DELTS=DELTH/TR-(1.D0-BR*VR*ZC/TR)+DLOG(TR/((VR-BR)*PR*ZC)) *
          ***********************************************************
                                      I
                                      I
                                      I
          ***********************************************************
I.........*      DO 3 I=1,4                                         *
I         ***********************************************************
I                                     I
I                                     I
I                                     I
I         ***********************************************************
I         *      VIRIAL(I)=FTC(I)+B(I)*(TR-1.D0)+C(I)*(ET-E)        *
I         ***********************************************************
I                                     I
I                                     I
I                                     I
I         ***********************************************************
..........* 3    DELTS=DELTS+VIRIAL(I)*FINV(I)/TR                   *
          ***********************************************************
                                      I
                                      I
                                      I
          ***********************************************************
I.........*      DO 4 J=1,2                                         *
I         ***********************************************************
I                                     I
I                                     I
I                                     I
I         ***********************************************************
I         *      VIRIAL(J+4)=FTC(J+4)+B(J+4)*(TR-1.D0)+C(J+4)*(ET-E) *
I         ***********************************************************
I                                     I

I                                     I
I                                     I
I                                     I
I                                     I
I         ***********************************************************
..........* 4    DELTS=DELTS+VIRIAL(J+4)*FINV(J+4)/TR               *
          ***********************************************************
                                      I
                                      I
                                      I
          ***********************************************************
          *      RETURN                                             *
          ***********************************************************

          ***********************************************************
          *      END                                                *
          ***********************************************************
```

```
                                        (ENTRANCE)
                                             I
                                             I
       ************************************************************************
       *      SUBROUTINE TOFSAP(TRA,TRB,PRA,PRB,SRA)                         *
       *      IMPLICIT REAL*8 (A-H,O-Z)                                      *
       *      DIMENSION FTC(6),B(6),C(6),CP(4)                               *
       *C                                                                    *
       *C       TOFSAP DETERMINES THE REDUCED TEMPERATURE AT STATE POINT A   *
       *C       GIVEN THE REDUCED PRESSURE AND REDUCED ENTROPY AT A.  THE    *
       *C       REDUCED PRESSURE  AND REDUCED TEMPERATURE AT SOME ARBITRARY  *
       *C       REFERENCE STATE B ARE REQUIRED TO ESTABLISH A REFERENCE CON- *
       *C       DITION.                                                      *
       *C                                                                    *
       *C       AN ITERATIVE  SOLUTION IS REQUIRED BECAUSE OF THE COMPLICATED *
       *C       TEMPERATURE DEPENDENCE OF ENTROPY FOR THE REAL GAS.  SUCCESSIVE *
       *C       APPROXIMATIONS TO TRA ARE MADE UNTIL THE DIFFERENCE BETWEEN THE *
       *C       CALCULATED AND GIVEN VALUES OF REDUCED ENTROPY IS SMALL.     *
       *C       THE REDUCED ENTROPY IS CALCULATED EACH TIME FROM THE RESULTS *
       *C       OF SUBROUTINES ENTRIG AND SDEV USING THE RELATIONSHIP:       *
       *C       REDUCED ENTROPY = REDUCED IDEAL GAS ENTROPY - REDUCED DEPARTURE *
       *C       FROM IDEAL.                                                  *
       *C                                                                    *
       *C       DEFINITION OF TERMS:                                         *
       *C       TRA = UNKNOWN REDUCED TEMPERATURE AT STATE POINT A.          *
       *C       TRB = REDUCED TEMPERATURE AT REFERENCE POINT B.              *
       *C       PRA = REDUCED PRESSURE AT STATE POINT A.                     *
       *C       PRB = REDUCED PRESSURE AT REFERENCE POINT B.                 *
       *C       SRA = REDUCED ENTROPY AT STATE POINT A.                      *
       *C                                                                    *
       *C                                                                    *
       *      CALL CSTNTS(AM,TC,VC,PC,ZC,FTC,B,C,AR,BR,RK,CP)                *
       *      CPIG=CP(1)                                                     *
       ************************************************************************
                                             I
                                             I
                                             I
       ************************************************************************
I......*      DO 1 I=2,4                                                     *
I      ************************************************************************
I                                            I
I                                            I
I                                            I
I      ************************************************************************
.......* 1    CPIG=CPIG+CP(I)*TRB**I                                         *
       ************************************************************************
                                             I
                                             I
                                             I
       ************************************************************************
       *      CALL VOLUME(VRB,TRB,PRB)                                       *
       *      CALL SDEV(DELTSB,TRB,PRB,VRB)                                  *
       *      CALL ENTRIG(TRB,PRB,SIGB)                                      *
       *      SRB=SIGB-DELTSB                                                *
       *      DUM1=TRB*(PRA/PRB)**(1.D0/CPIG)                                *
       *      DUM2=DUM1*DEXP((SRA-SRB)/CPIG)                                 *
       ************************************************************************
                                             I


                                             I
                                             I
                                             I
                                             I
       ************************************************************************
       *      IF(PRA.GT.1.D0)GO TO 2                                         *......O
       ************************************************************************      I
                                             I                                      I
                                             I                                      I
                                             I                                      I
       ************************************************************************      I
       *      CALL TRSAT(PRA,DUM3)                                           *      I
       ************************************************************************      I
                                             I                                      I
                                             O(.....................................O
                                             I
       ************************************************************************
       * 2    IF(PRA.GT.1.D0)DUM3=.975D0                                     *
       ************************************************************************
                                             I
                                             I
                                             I
       ************************************************************************
       *      IF((SRA-SRB).GT.0.) GO TO 3                                    *......O
       ************************************************************************      I
                                             I                                      I
```

```
*       IF((SRA-SRB).LE.0.) GO TO 5

* 3     TU=DUM2+.05D0
*       TI=DUM3
*       TMEAN=(TL+TU)/2.D0

* 4     CALL VOLUME(VMEAN,TMEAN,PRA)
*       CALL SDEV(DELTSM,TMEAN,PRA,VMEAN)
*       CALL ENTRIG(TMEAN,PRA,SIGM)
*       SRMEAN=SIGM-DELTSM
*       DEL=(SRMEAN-SRA)/DABS(SRA)

*       IF(DABS(DEL).LE..5D-3) GO TO 7
```

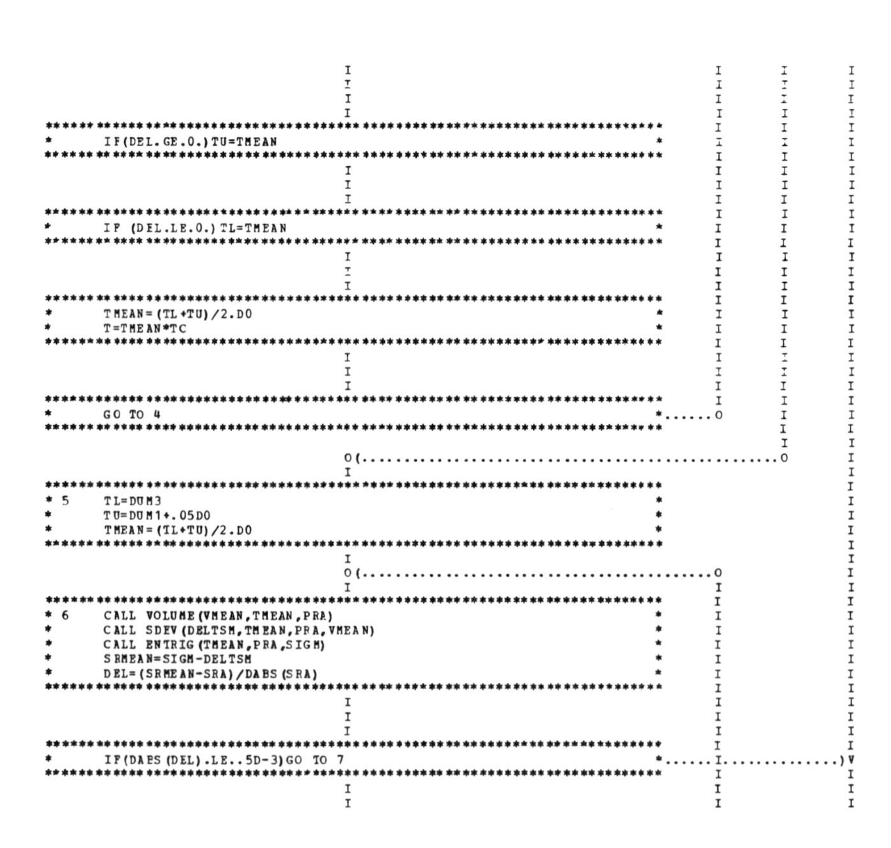

```
*       IF(DEL.GE.0.)TU=TMEAN

*       IF (DEL.LE.0.) TL=TMEAN

*       TMEAN=(TL+TU)/2.D0
*       T=TMEAN*TC

*       GO TO 4

* 5     TL=DUM3
*       TU=DUM1+.05D0
*       TMEAN=(TL+TU)/2.D0

* 6     CALL VOLUME(VMEAN,TMEAN,PRA)
*       CALL SDEV(DELTSM,TMEAN,PRA,VMEAN)
*       CALL ENTRIG(TMEAN,PRA,SIGM)
*       SRMEAN=SIGM-DELTSM
*       DEL=(SRMEAN-SRA)/DABS(SRA)

*       IF(DABS(DEL).LE..5D-3)GO TO 7
```

```
*      IF(DEL.GE.0.)TU=TMEAN

*      IF(DEL.LE.0.)TL=TMEAN

*      TMEAN=(TL+TU)/2.D0

*      GO TO 6

* 7    TRA=TMEAN

*      RETURN

*      END

                          (ENTRANCE)

*      SUBROUTINE TOPHAP(TRA,TRB,PRA,PRB,HRA)
*      IMPLICIT REAL*8 (A-H,O-Z)
*      DIMENSION FTC(6),B(6),C(6),CP(4)
*C
*C       TOPHAP DETERMINES THE REDUCED TEMPERATURE AT STATE POINT A
*C       GIVEN THE REDUCED PRESSURE AND REDUCED ENTHALPY AT A. THE
*C       REDUCED PRESSURE AND REDUCED TEMPERATURE AT SOME ARBITRARY
*C       REFERENCE STATE B ARE REQUIRED TO ESTABLISH A REFERENCE CON-
*C       DITION.
*C
*C       LIKE TOPSAP, AN ITERATIVE SOLUTION IS USED.
*C
*C       DEFINITION OF TERMS:
*C       TRA = UNKNOWN REDUCED TEMPERATURE AT STATE POINT A.
*C       TRB = REDUCED TEMPERATURE AT REFERENCE POINT B.
*C       PRA = REDUCED PRESSURE AT STATE POINT A.
*C       PRB = REDUCED PRESSURE AT REFERENCE POINT B.
*C       HRA = REDUCED ENTHALPY AT STATE POINT A.
*C
*C       SUCCESSIVE APPROXIMATIONS TO TRA ARE MADE UNTIL THE DIFFERENCE
*C       BETWEEN THE CALCULATED AND GIVEN VALUES OF REDUCED ENTHALPY IS
*C       SMALL.  THE REDUCED ENTHALPY IS CALCULATED EACH TIME FROM THE
*C       RESULTS OF SUBROUTINES ENTHIG AND HDEV USING THE RELATIONSHIP:
*C       REDUCED ENTHALPY = REDUCED IDEAL GAS ENTHALPY - REDUCED DEPAR-
*C       TURE FROM IDEAL.
*C
*      CALL CSTNTS(AM,TC,VC,PC,ZC,FTC,B,C,AR,BR,RK,CP)
*      CPIG=CP(1)

*      DO 1 I=2,4

* 1    CPIG=CPIG+CP(I)*TRB**I
```

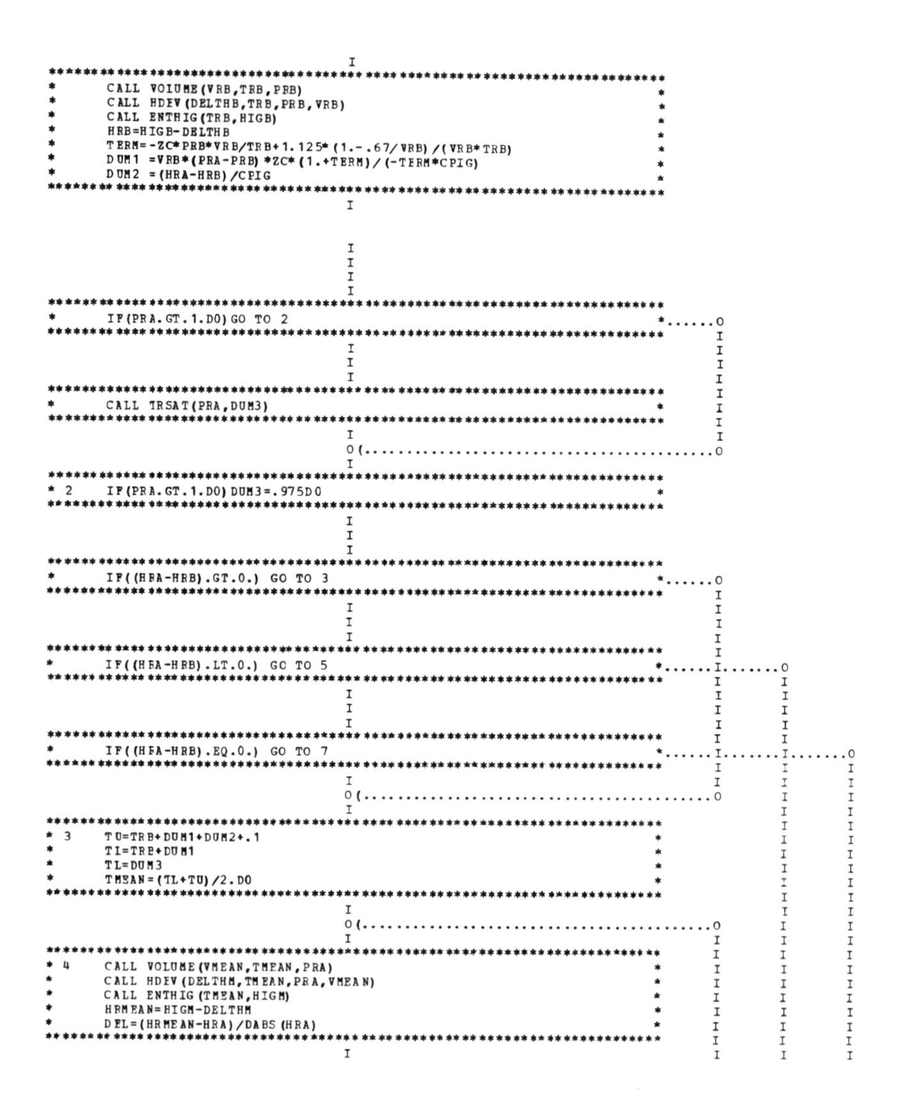

```
      CALL VOLUME(VRB,TRB,PRB)
      CALL HDEV(DELTHB,TRB,PRB,VRB)
      CALL ENTHIG(TRB,HIGB)
      HRB=HIGB-DELTHB
      TERM=-ZC*PRB*VRB/TRB+1.125*(1.-.67/VRB)/(VRB*TRB)
      DUM1 =VRB*(PRA-PRB)*ZC*(1.+TERM)/(-TERM*CPIG)
      DUM2 =(HRA-HRB)/CPIG

      IF(PRA.GT.1.D0)GO TO 2

      CALL TRSAT(PRA,DUM3)

2     IF(PRA.GT.1.D0)DUM3=.975D0

      IF((HRA-HRB).GT.0.) GO TO 3

      IF((HRA-HRB).LT.0.) GO TO 5

      IF((HRA-HRB).EQ.0.) GO TO 7

3     TU=TRB+DUM1+DUM2+.1
      TI=TRB+DUM1
      TL=DUM3
      TMEAN=(TL+TU)/2.D0

4     CALL VOLUME(VMEAN,TMEAN,PRA)
      CALL HDEV(DELTHM,TMEAN,PRA,VMEAN)
      CALL ENTHIG(TMEAN,HIGM)
      HRMEAN=HIGM-DELTHM
      DEL=(HRMEAN-HRA)/DABS(HRA)
```

```
*      IF(DABS(DEL).LE..5D-3) GO TO 9

*      IF(DEL.GE.0.)TU=TMEAN

*      IF (DEL.LE.0.) TL=TMEAN

*      TMEAN=(TL+TU)/2.D0

*      GO TO 4

* 5    TL=TRB+DUM1+DUM2
*      TI=DUM3
*      TU=TRB+DUM1+.1
*      TMEAN=(TL+TU)/2.D0

* 6    CALL VOLUME(VMEAN,TMEAN,PRA)
*      CALL HDEV(DELTHM,TMEAN,PRA,VMEAN)
*      CALL ENTHIG(TMEAN,HIGM)
*      HRMEAN=HIGM-DELTHM
*      DEL=(HRMEAN-HRA)/DABS(HRA)

*      IF(DABS(DEL).LE..5D-3)GO TO 9
```

```
*      IF(DEL.GE.0.)TU=TMEAN

*      IF(DEL.LE.0.)TL=TMEAN

*      TMEAN=(TL+TU)/2.D0

*      GO TO 6
```

```
* 7     TL=TRB
*       TU=TRB+DUM1+.1

*       IF(PRA-PRB.LE.0.)TL=TRB+DUM1

*       IF(PRA-PRB.LE.0.)TU=TRB

*       TL=DUM3
*       TMEAN=(TL+TU)/2.D0

* 8     CALL VOLUME(VMEAN,TMEAN,PRA)
*       CALL HDEV(DELTHM,TMEAN,PRA,VMEAN)
*       CALL ENTHIG(TMEAN,HIGM)
*       HRMEAN=HIGM-DELTHM
*       DEL=(HRMEAN-HRA)/DABS(HRA)

*       IF(DABS(DEL).LE..5D-3)GO TO 9

*       IF(DEL.GE.0.)TU=TMEAN

*       IF(DEL.LE.0.)TL=TMEAN

*       TMEAN=(TL+TU)/2.D0

*       GO TO 8

* 9     TRA=TMEAN

*       RETURN

*       END
```

```
                                    (ENTRANCE)
                                        I
                                        I
***************************************************************************
*       SUBROUTINE ENTHAL(T,P,H)                                          *
*       IMPLICIT REAL*8(A-H,O-Z)                                          *
*       DIMENSION FTC(6),B(6),C(6),CP(4)                                  *
*       CALL CSTNTS(AM,TC,VC,PC,ZC,FTC,B,C,AR,BR,RK,CP)                   *
*C                                                                        *
*C        ENTHAL COMPUTES THE ENTHALPY OF THE REAL GAS ABOVE THE SATURATED*
*C        LIQUID VALUE AT -40F IN BTU/LB, GIVEN THE TEMPERATURE IN DEG. R *
*C        AND PRESSURE IN PSIA.                                           *
*C                                                                        *
*C        SUBROUTINE  ZERO IS CALLED TO DETERMINE THE REFERENCE STATE.    *
*C                                                                        *
*C        DEFINITION OF TERMS:                                            *
*C        T = ABSOLUTE TEMPERATURE IN DEGREES RANKINE.                    *
*C        P = ABSOLUTE PRESSURE IN PSIA.                                  *
*C        H = ENTHALPY IN BTU/LBM.                                        *
*C                                                                        *
*       CALL ZERO(TO,PSO,VO,HFGO)                                         *
*       TRO=TO/TC                                                         *
*       PSRO=PSO/(14.696D0*PC)                                            *
*       HFGRO=HFGO*AM/(1.987D0*TC)                                        *
*       CALL VOLUME(VRO,TRO,PSRO)                                         *
*       CALL HDEV(DELTHO,TRO,PSRO,VRO)                                    *
*       CALL ENTHIG(TRO,HIGO)                                             *
*       HSLO=(HIGO-DELTHO)*1.987*TC/AM-HFGO                               *
*       TR=T/TC                                                           *
*       PR=P/(PC*14.696D0)                                                *
*       CALL VOLUME(VR,TR,PR)                                             *
*       CALL HDEV(DELTH,TR,PR,VR)                                         *
*       CALL ENTHIG(TR,HIG)                                               *
*       H=(HIG-DELTH)*1.987*TC/AM-HSLO                                    *
***************************************************************************
                                        I
                                        I
                                        I
***************************************************************************
*       RETURN                                                            *
***************************************************************************


***************************************************************************
*       END                                                               *
***************************************************************************

                                    (ENTRANCE)
                                        I
                                        I
***************************************************************************
*       SUBROUTINE ENTROP(T,P,S)                                          *
*       IMPLICIT REAL*8(A-H,O-Z)                                          *
*       DIMENSION FTC(6),B(6),C(6),CP(4)                                  *
*C                                                                        *
*C        ENTROP COMPUTES THE ENTROPY OF THE REAL GAS ABOVE THE SATURATED *
*C        LIQUID VALUE AT -40F IN BTU/LB-DEG.R, GIVEN THE TEMPERATURE IN  *
*C        DEGREES RANKINE AND THE PRESSURE IN PSIA.                       *
*C                                                                        *
*C        SUBROUTINE ZERO IS CALLED TO DETERMINE THE REFERENCE STATE.     *
*C                                                                        *
*C        DEFINITION OF TERMS:                                            *
*C        T = ABSOLUTE TEMPERATURE IN DEGREES RANKINE.                    *
*C        P = ABSOLUTE PRESSURE IN PSIA.                                  *
*C        S = ENTROPY IN BTU/LBM.-DEG.R.                                  *
*C                                                                        *
*       CALL CSTNTS(AM,TC,VC,PC,ZC,FTC,B,C,AR,BR,RK,CP)                   *
*       CALL ZERO(TO,PSO,VO,HFGO)                                         *
*       TRO=TO/TC                                                         *
*       PSRO=PSO/(14.696D0*PC)                                            *
*       HFGRO=HFGO*AM/(1.987D0*TC)                                        *
*       CALL VOLUME(VRO,TRO,PSRO)                                         *
*       CALL SDEV(DELTSO,TRO,PSRO,VRO)                                    *
*       CALL ENTRIG(TRO,PSRO,SIGO)                                        *
*       SSLO=(SIGO-DELTSO)*1.987/AM-HFGO/TO                               *
*       TR=T/TC                                                           *
*       PR=P/(PC*14.696D0)                                                *
*       CALL VOLUME(VR,TR,PR)                                             *
*       CALL SDEV(DELTS,TR,PR,VR)                                         *
*       CALL ENTRIG(TR,PR,SIG)                                            *
*       S=(SIG-DELTS)*1.987/AM-SSLO                                       *
***************************************************************************
                                        I
                                        I
                                        I
***************************************************************************
*       RETURN                                                            *
***************************************************************************


***************************************************************************
*       END                                                               *
***************************************************************************
```

```
                                 (ENTRANCE)
                                     I
                                     I
***************************************************************************
*       SUBROUTINE TEMPH(H,P,T)                                           *
*       IMPLICIT REAL*8 (A-H,O-Z)                                         *
*       DIMENSION FTC(6),B(6),C(6),CP(4)                                  *
*C                                                                        *
*C        TEMPH  COMPUTES THE TEMPERATURE IN DEGREES RANKINE GIVEN THE    *
*C        PRESSURE IN PSIA, AND THE ENTHALPY ABOVE THE SATURATED LIQUID   *
*C        AT -40F IN BTU/LBM.                                             *
*C                                                                        *
*C        SUBROUTINE ZERO IS CALLED TO ESTABLISH THE REFERENCE CONDITIONS.*
*C        THE THERMODYNAMIC PROPERTIES ARE REDUCED TO THE FORMAT REQUIRED *
*C        BY SUBROUTINE TOPHAP WHICH PERFORMS THE TEMPERATURE/ENTHALPY(RE-*
*C        DUCED) ITERATIVE SOLUTION.                                      *
*C                                                                        *
*C        DEFINITION OF TERMS:                                            *
*C        H = SPECIFIC ENTHALPY IN BTU/LBM.                               *
*C        P = ABSOLUTE PRESSURE IN PSIA.                                  *
*C        T = ABSOLUTE TEMPERATURE IN DEGREES RANKINE.                    *
*C                                                                        *
*       CALL CSTNTS(AM,TC,VC,PC,ZC,FTC,B,C,AR,BR,RK,CP)                   *
*       CALL ZERO(TO,PSO,VO,HFGO)                                         *
*       TRO=TO/TC                                                         *
*       PSRO=PSO/(14.696D0*PC)                                            *
*       HFGRO=HFGO*AM/(1.987D0*TC)                                        *
*       CALL VOLUME(VRO,TRO,PSRO)                                         *
*       CALL HDEV(DELTHO,TRO,PSRO,VRO)                                    *
*       CALL ENTHIG(TRO,HIGO)                                             *
*       HSLO=(HIGO-DELTHO)*1.987*TC/AM-HFGO                               *
*       PR=P/(PC*14.696D0)                                                *
*       HR=(H+HSLO)*AM/(TC*1.987D0)                                       *
*       CALL TOPHAP(TR,TRO,PR,PSRO,HR)                                    *
*       T=TR*TC                                                           *
***************************************************************************
                                     I
                                     I
                                     I
***************************************************************************
*       RETURN                                                            *
***************************************************************************


***************************************************************************
*       END                                                               *
***************************************************************************
                                 (ENTRANCE)
                                     I
                                     I
***************************************************************************
*       SUBROUTINE TEMPS(S,P,T)                                           *
*       IMPLICIT REAL*8 (A-H,O-Z)                                         *
*       DIMENSION FTC(6),B(6),C(6),CP(4)                                  *
*C                                                                        *
*C        TEMPS COMPUTES THE TEMPERATURE IN DEGREES RANKINE GIVEN THE     *
*C        PRESSURE IN PSIA. AND THE ENTROPY ABOVE THE SATURATED LIQUID    *
*C        AT -40F IN BTU/LBM.-DEG.R.                                      *
*C                                                                        *
*C        SUBROUTINE ZERO IS CALLED TO ESTABLISH THE REFERENCE CONDITIONS.*
*C        THE THERMODYNAMIC PROPERTIES ARE REDUCED TO THE FORMAT REQUIRED *
*C        BY SUBROUTINE TOPSAP, WHICH PERFORMS THE TEMPERATURE/ENTROPY(RE-*
*C        DUCED) ITERATIVE SOLUTION.                                      *
*C                                                                        *
*C        DEFINITION OF TERMS:                                            *
*C        S = SPECIFIC ENTROPY IN BTU/LBM.-DEG.R.                         *
*C        P = ABSOLUTE PRESSURE IN PSIA.                                  *
*C        T = ABSOLUTE TEMPERATURE IN DEGREES RANKINE.                    *
*C                                                                        *
*       CALL CSTNTS(AM,TC,VC,PC,ZC,FTC,B,C,AR,BR,RK,CP)                   *
*       CALL ZERO(TO,PSO,VO,HFGO)                                         *
*       TRO=TO/TC                                                         *
*       PSRO=PSO/(14.696D0*PC)                                            *
*       HFGRO=HFGO*AM/(1.987D0*TC)                                        *
*       CALL VOLUME(VRO,TRO,PSRO)                                         *
*       CALL SDEV(DELTSO,TRO,PSRO,VRO)                                    *
*       CALL ENTRIG(TRO,PSRO,SIGO)                                        *
*       SSLO=(SIGO-DELTSO)*1.987/AM-HFGO/TO                               *
*       SR=(S+SSLO)*AM/(1.987D0)                                          *
*       PR=P/(PC*14.696D0)                                                *
*       CALL TOPSAP(TR,TRO,PR,PSRO,SR)                                    *
*       T=TR*TC                                                           *
***************************************************************************
                                     I
                                     I
                                     I
***************************************************************************
*       RETURN                                                            *
***************************************************************************


***************************************************************************
*       END                                                               *
***************************************************************************
```

```
                                  (ENTRANCE)
                                      I
                                      I
*****************************************************************************
*       SUBROUTINE HOFVT(HR,VR,TR)                                          *
*       IMPLICIT REAL*8(A-H,O-Z)                                            *
*       DIMENSION FTC(6),B(6),C(6),CP(4)                                    *
*C                                                                          *
*C        HOFVT COMPUTES THE REDUCED ENTHALPY ABOVE THE SATURATED LIQUID    *
*C        AT -40F GIVEN THE REDUCED VOLUME VR AND THE REDUCED TEMPERATURE   *
*C        TR.                                                               *
*C                                                                          *
*C        DEFINITION OF TERMS:                                              *
*C        HR = REDUCED ENTHALPY(DIMENSICNLESS).                             *
*C        HR = ENTHALPY/(CRITICAL TEMPERATURE*GAS CONSTANT).                *
*C        VR = REDUCED VOLUME.                                              *
*C        TR = REDUCED TEMPERATURE.                                         *
*C                                                                          *
*       CALL CSTNTS(AM,TC,VC,PC,ZC,FTC,B,C,AR,BR,RK,CP)                     *
*       CALL ZERO(TO,PSO,VO,HFGO)                                           *
*       TRO=TO/TC                                                           *
*       PSRO=PSO/(14.696D0*PC)                                              *
*       VRO=VO/VC                                                           *
*       HFGRO=HFGO*AM/(1.987D0*TC)                                          *
*       CALL HDEV(DELTHO,TRO,PSRO,VRO)                                      *
*       CALL ENTHIG(TRO,HIGO)                                               *
*       HSLRO=HIGO-DELTHO-HFGRO                                             *
*       CALL PRES(PR,TR,VR)                                                 *
*       CALL HDEV(DELTH,TR,PR,VR)                                           *
*       CALL ENTHIG(TR,HIG)                                                 *
*       HR=HIG-DELTH-HSLRO                                                  *
*****************************************************************************
                                      I
                                      I
                                      I
*****************************************************************************
*       RETURN                                                              *
*****************************************************************************


*****************************************************************************
*       END                                                                 *
*****************************************************************************

                                  (ENTRANCE)
                                      I
                                      I
*****************************************************************************
*       SUBROUTINE SOFVT(SR,VR,TR)                                          *
*       IMPLICIT REAL*8(A-H,O-Z)                                            *
*       DIMENSION FTC(6),B(6),C(6),CP(4)                                    *
*C                                                                          *
*C        SOFVT COMPUTES THE REDUCED ENTROPY SR ABOVE THE SATURATED LIQUID*
*C        AT -40F GIVEN THE REDUCED VOLUMF VR AND THE REDUCED TEMPERATURE   *
*C        TR.                                                               *
*C                                                                          *
*C        DEFINITION OF TERMS:                                              *
*C        SR = REDUCED ENTROPY(DIMENSIONLESS).                              *
*C        SR = ENTROPY/GAS CONSTANT.                                        *
*C        VR = REDUCED VOLUME.                                              *
*C        TR = REDUCED TEMPERATURE.                                         *
*C                                                                          *
*       CALL CSTNTS(AM,TC,VC,PC,ZC,FTC,B,C,AR,BR,RK,CP)                     *
*       CALL ZERO(TO,PSO,VO,HFGO)                                           *
*       TRO=TO/TC                                                           *
*       PSRO=PSO/(14.696D0*PC)                                              *
*       VRO=VO/VC                                                           *
*       HFGRO=HFGO*AM/(1.987D0*TC)                                          *
*       CALL SDEV(DELTSO,TRO,PSRO,VRO)                                      *
*       CALL ENTRIG(TRO,PSRO,SIGO)                                          *
*       SSLRO=SIGO-DELTSO-HFGRO/TRO                                         *
*       CALL PRES(PR,TR,VR)                                                 *
*       CALL SDEV(DELTS,TR,PR,VR)                                           *
*       CALL ENTRIG(TR,PR,SIG)                                              *
*       SR=SIG-DELTS-SSLRO                                                  *
*****************************************************************************
                                      I
                                      I
                                      I
*****************************************************************************
*       RETURN                                                              *
*****************************************************************************


*****************************************************************************
*       END                                                                 *
*****************************************************************************
```

```
                                  (ENTRANCE)
                                      I
                                      I
************************************************************************
*      SUBROUTINE PRES(PR,TR,VR)
*      IMPLICIT REAL*8 (A-H,O-Z)
*      DIMENSION FTC(6),B(6),C(6),CP(4)
*      DIMENSION F(6)
*C
*C       PRES COMPUTES THE REDUCED PRESSURE PR CORRESPONDING TO THE
*C       REDUCED TEMPERATURE TR AND REDUCED VOLUME VR.
*C
*C       DEFINITION OF TERMS:
*C       PR = REDUCED PRESSURE.
*C       TR = REDUCED TEMPERATURE.
*C       VR = REDUCED VOLUME.
*C
*      CALL CSTNTS(AM,TC,VC,PC,ZC,FTC,B,C,AR,BR,RK,CP)
************************************************************************
                                      I
                                      I
                                      I
************************************************************************
I...................*      DO 1 M=1,6
I                   ****************************************************
I                                     I
I                                     I
I                                     I
I                   ****************************************************
....................* 1    F(M)=FTC(M)+B(M)*(TR-1.D0)+C(M)*(DEXP(-TR*RK)-DEXP(-RK))
                    ****************************************************
                                      I
                                      I
                                      I
************************************************************************
*      TERM=1.D0/(VR-BR)
*      PR =(1.D0/ZC)*TERM*(TR+TERM*(F(1)+TERM*(F(2)+TERM*
*     1(F(3)+TERM*(F(4))))))+(1.D0/ZC)*DEXP(-AR*VR)*(F(5)+F
*     2(6)*DEXP(-AR*VR))
************************************************************************
                                      I
                                      I
                                      I
************************************************************************
*      RETURN
************************************************************************


************************************************************************
*      END
************************************************************************

                                  (ENTRANCE)
                                      I
                                      I
************************************************************************
*      SUBROUTINE ZERO(TO,PSO,VO,HFGO)
*      IMPLICIT REAL*8 (A-H,O-Z)
*      DIMENSION FTC(6),B(6),C(6),CP(4)
*C
*C       SUBROUTINE ZERO PROVIDES THE TEMPERATURE, PRESSURE, SPECIFIC
*C       VOLUME, AND ENTHALPY OF THE SATURATED VAPOR ABOVE THE REFERENCE
*C       STATE(SATURATED LIQUID AT -40 F).  ENGINEERING UNITS ARE USED.
*C
*C       DEFINITION OF TERMS:
*C       TO = REFERENCE TEMPERATURE IN DEGREES RANKINE.
*C       PSO = SATURATION PRESSURE AT TO IN PSIA.
*C       VO  = SPECIFIC VOLUME OF SATURATED VAPOR AT TO.
*C       HFGO = LATENT HEAT OF VAPORIZATION AT TO IN BTU/LBM.
*C
*      CALL CSTNTS(AM,TC,VC,PC,ZC,FTC,B,C,AR,BR,RK,CP)
*      TO=419.58D0
*      TRO=TO/TC
*      CALL PRSAT(TRO,PRSO)
*      PSO=PRSO*PC*14.696D0
*      CALL VOLUME(VRO,TRO,PRSO)
*      VO=VRO*VC
*      CALL LATENT(TRO,HFGRO)
*      HFGO=HFGRO*1.987D0*TC/AM
************************************************************************
                                      I
                                      I
                                      I
************************************************************************
*      RETURN
************************************************************************


************************************************************************
*      END
************************************************************************
```

```
                                   (ENTRANCE)
                                        I
                                        I
*******************************************************************************
*       SUBROUTINE PRSAT(TR,PR)                                               *
*       IMPLICIT REAL*8 (A-H,O-Z)                                             *
*C                                                                            *
*C        PRSAT COMPUTES THE REDUCED SATURATION PRESSURE PR AT THE            *
*C        REDUCED TEMPERATURE TR.                                             *
*C                                                                            *
*C        DEFINITION OF TERMS:                                                *
*C        TR = REDUCED TEMPERATURE.                                           *
*C        PR = REDUCED SATURATION PRESSURE.                                   *
*C        X0,X1,X2,X3,X4 = COEFFICIENTS FOR SATURATION PRESSURE EQUATION.     *
*C        THESE COEFFICIENTS WILL DIFFER FOR DIFFERENT COMPOUNDS.             *
*C                                                                            *
*       X0=7.3672454D0                                                        *
*       X1=10.8839450D0                                                       *
*       X2=8.5673754D0                                                        *
*       X3=-2.4647451D0                                                       *
*       X4=1.0519545D0                                                        *
*       X=X0-X1/TR-X2*DLOG(TR)-X3*TR+X4*TR*TR                                 *
*       PR=DEXP(X)                                                            *
*******************************************************************************
                                        I
                                        I
                                        I
*******************************************************************************
*       RETURN                                                                *
*******************************************************************************


*******************************************************************************
*       END                                                                   *
*******************************************************************************
```

```
                                   (ENTRANCE)
                                        I
                                        I
*******************************************************************************
*       SUBROUTINE DPRSAT(TR,DPR)                                             *
*       IMPLICIT REAL*8 (A-H,O-Z)                                             *
*C                                                                            *
*C        DPRSAT COMPUTES THE REDUCED SLOPE OF THE VAPOR PRESSURE CURVE,      *
*C        AT REDUCED TEMPERATURE TR.  THIS IS NEEDED TO DETERMINE THE         *
*C        LATENT HEAT OF VAPORIZATION.                                        *
*C        DEFINITION OF TERMS.                                                *
*C        TR = REDUCED TEMPERATURE(MUST BE LESS THAN OR EQUAL TO 1.0)         *
*C        DPR = SLOPE OF REDUCED VAPOR PRESSURE CURVE.                        *
*C        X0,X1,X2,X3,X4 = COEFFICIENTS FOR SATURATION PRESSURE EQUATION.     *
*C        THESE COEFFICIENTS ARE IDENTICAL TO THOSE OF SUBROUTINE PRSAT.      *
*C                                                                            *
*       X0=9.2853003D0                                                        *
*       X1=10.39641863D0                                                      *
*       X2=6.4012471D0                                                        *
*       X3=.6970878097D0                                                      *
*       X4=1.810124278D0                                                      *
*       X=X0-X1/TR-X2*DLOG(TR)-X3*TR+X4*TR*TR                                 *
*       PR=DEXP(X)                                                            *
*       DPR=PR*(X1/TR**2-X2/TR-X3+2.*X4*TR)                                   *
*******************************************************************************
                                        I
                                        I
                                        I
*******************************************************************************
*       RETURN                                                                *
*******************************************************************************


*******************************************************************************
*       END                                                                   *
*******************************************************************************
```

```
                                   (ENTRANCE)
                                       I
                                       I
*****************************************************************************
*       SUBROUTINE LATENT(TR,HFGR)                                          *
*       IMPLICIT REAL*8(A-H,O-Z)                                            *
*       DIMENSION FTC(6),B(6),C(6),CP(4)                                    *
*C                                                                          *
*C        LATENT COMPUTES THE VALUE OF REDUCED LATENT HEAT HFGR, GIVEN      *
*C        THE REDUCED TEMPERATURE TR.                                       *
*C                                                                          *
*C        DEFINITION OF TERMS:                                              *
*C        TR = REDUCED TEMPERATURE(MUST BE LESS THAN OR EQUAL TO 1.0).      *
*C        HFGR = REDUCED LATENT HEAT.                                       *
*C        HFGR = LATENT HEAT/(CRITICAL TEMPERATURE*GAS CONSTANT).           *
*C                                                                          *
*       CALL CSTNTS(AM,TC,VC,PC,ZC,FTC,B,C,AR,BR,RK,CP)                     *
*       CALL PRSAT(TR,PR)                                                   *
*       CALL DPRSAT(TR,DPR)                                                 *
*       CALL LQDENS(TR,RHOR)                                                *
*       CALL VOLUME(VR,TR,PR)                                               *
*       HFGR=ZC*TR*DPR*(VR-1./RHOR)                                         *
*****************************************************************************
                                       I
                                       I
                                       I
*****************************************************************************
*       RETURN                                                              *
*****************************************************************************


*****************************************************************************
*       END                                                                 *
*****************************************************************************

                                   (ENTRANCE)
                                       I
                                       I
*****************************************************************************
*       SUBROUTINE LQDENS(TR,RHOR)                                          *
*       IMPLICIT REAL*8(A-H,O-Z)                                            *
*C                                                                          *
*C        LODENS COMPUTES THE REDUCED SATURATED LIQUID DENSITY RHOR, GIVEN*
*C        THE REDUCED TEMPERATURE TR.                                       *
*C                                                                          *
*C        DEFINITION OF TERMS:                                              *
*C        TR = REDUCED TEMPERATURE(MUST BE LESS THAN OR EQUAL TO 1.0).      *
*C        RHOR = REDUCED SATURATED LIQUID DENSITY.                          *
*C        RHOR = DENSITY*CRITICAL SPECIFIC VOLUME.                          *
*C        A,B,C,D = LIQUID DENSITY EQUATION COEFFICIENTS(DIMENSIONLESS).    *
*C                                                                          *
*       A=1.5489388D0                                                       *
*       B=1.9710952D0                                                       *
*       C=-1.5957075D0                                                      *
*       D=1.0321973D0                                                       *
*       X=0.D0                                                              *
*****************************************************************************
                                       I
                                       I
                                       I
*****************************************************************************
*       IF(DABS(1.-TR).LT.1.D-6) GO TO 10                                   *......O
*****************************************************************************      I
                                       I                                           I
                                       I                                           I
                                       I                                           I
*****************************************************************************      I
*       X=(1.-TR)**(1./3.)                                                  *      I
*****************************************************************************      I
                                       I                                           I
                                       O(..........................................O
                                       I
*****************************************************************************
*    10 CONTINUE                                                            *
*       RHOR=1.0+A*X+B*X*X+C*X*X*X+D*X*X*X*X                                *
*****************************************************************************
                                       I
                                       I
                                       I
*****************************************************************************
*       RETURN                                                              *
*****************************************************************************


*****************************************************************************
*       END                                                                 *
*****************************************************************************
```

```
(ENTRANCE)

      SUBROUTINE TRSAT(PR,TR)
      IMPLICIT REAL*8 (A-H,O-Z)
C
C       COMPUTES REDUCED SATURATION TEMPERATURE TR GIVEN
C       REDUCED PRESSURE PR.
C
      TRL=.4D0
      TRU=1.01D0
      TMEAN=(TRL+TRU)/2.D0

1     TR=TMEAN
      CALL PRSAT(TR,PRMEAN)
      DEL=(PRMEAN-PR)/PR

      IF(DABS(DEL).LE.1.D-6)GO TO 2

      IF(DEL.GE.0.)TRU=TMEAN

      IF(DEL.LE.0.)TRL=TMEAN

      TMEAN=(TRU+TRL)/2.D0

      GO TO 1

2     TR=TMEAN

      RETURN

      END
```

```
                              (ENTRANCE)
                                  I
                                  I
*************************************************************************
*       SUBROUTINE DTRDH(DELHR,TR1,TR2)                                 *
*       IMPLICIT REAL*8(A-H,O-Z)                                        *
*C         COMPUTES LIQUID REDUCED TEMPERATURE CHANGE GIVEN             *
*C         REDUCED ENTHALPY CHANGE AT CONSTANT PRESSURE.                *
*C         TR2=TR2(HR1+DELHR) WHERE HR1 IS SATURATED LIQUID             *
*C         ENTHALPY AT TR1.  TR2 IS TEMPERATURE FOR WHICH SAT-          *
*C         URATED LIQUID ENTHALPY IS HR1+DELHR.                         *
*       CALL PRSAT(TR1,PR1)                                             *
*       CALL VOLUME(VR1,TR1,PR1)                                        *
*       CALL LATENT(TR1,HFGR1)                                          *
*       CALL HOFVT(HRSV1,VR1,TR1)                                       *
*       HR1=HRSV1-HFGR1                                                 *
*       HR=HR1+DELHR                                                    *
*       TRL=TR1-.1D0                                                    *
*       TRU=1.D0                                                        *
*       TMEAN=(TRU+TRL)/2.D0                                            *
* 1     CONTINUE                                                        *
*************************************************************************
                                  I
                                  O(........................................O
                                  I                                         I
*************************************************************************   I
* 2     TR=TMEAN                                                        *   I
*       CALL PRSAT(TR,PR)                                               *   I
*       CALL VOLUME(VR,TR,PR)                                           *   I
*       CALL LATENT(TR,HFGR)                                            *   I
*       CALL HOFVT(HRSV,VR,TR)                                          *   I
*       HRMEAN=HRSV-HFGR                                                *   I
*       DIFF=(HRMEAN-HR)/HR                                             *   I
*************************************************************************   I
                                  I                                         I
                                  I                                         I
                                  I                                         I
*************************************************************************   I
*       IF(DABS(DIFF).LE.1.D-3)GO TO 3                                  *...I.......O
*************************************************************************   I       I
                                  I                                         I       I
                                  I                                         I       I
                                  I                                         I       I
*************************************************************************   I       I
*       IF(DIFF.GE.0.)TRU=TMEAN                                         *   I       I
*************************************************************************   I       I
                                  I                                         I       I
                                  I                                         I       I
                                  I                                         I       I
*************************************************************************   I       I
*       IF(DIFF.LE.0.)TRL=TMEAN                                         *   I       I
*************************************************************************   I       I
                                  I                                         I       I
                                  I                                         I       I
                                  I                                         I       I


                                  I                                         I       I
                                  I                                         I       I
*************************************************************************   I       I
*       TMEAN=(TRL+TRU)/2.D0                                            *   I       I
*************************************************************************   I       I
                                  I                                         I       I
                                  I                                         I       I
                                  I                                         I       I
*************************************************************************   I       I
*       GO TO 2                                                         *...O       I
*************************************************************************           I
                                                                                    I
                                  O(................................................O
                                  I
*************************************************************************
* 3     TR2=TR                                                          *
*************************************************************************
                                  I
                                  I
                                  I
*************************************************************************
*       RETURN                                                          *
*************************************************************************


*************************************************************************
*       END                                                             *
*************************************************************************
```

9. NOMENCLATURE

A — surface area for heat transport.

a, A_i — constants in the Martin-Hou equation of state.

A,B,C,D, — empirical coefficients in the reduced liquid density equation.

A_r, B_r, C_r, D_r, E_r — empirical coefficients in the reduced vapor pressure equation.

b, B_i, C_i — constants in the Martin-Hou equation of state.

$C_o^*, C_1^*, C_2^*, C_3^*$ — empirical coefficients in the heat capacity equation.

C_p — heat capacity at constant pressure.

C_p^* — ideal gas state heat capacity at constant pressure.

C_v — heat capacity at constant volume.

D — equivalent hydraulic diameter.

D_p — turbine pitch diameter.

D_s — specific turbine diameter.

$E_r(T_r)$ — dimensionless volumetric energy density.

f_1 — correction factor for turbine blade tip speed effects on turbine cost.

f_2 — correction factor for pressure on turbine cost.

f_I, f_i — cost fraction factors.

$f_i(T_r)$ — parameter in the Martin-Hou equation of state.

g — acceleration of gravity.

H — enthalpy.

h — heat transfer coefficient.

h^* — open flow length on the turbine blade.

h_{fg} — enthalpy of vaporization.

I — component irreversibility.

K — constant in the Martin-Hou equation of state.

k — thermal conductivity.

L — length of heating or condensing section.

$\dot{m}$ — mass flow rate.

$\mathcal{M}$ — molecular weight.

N — turbine rotational speed.

N_s — specific turbine speed.

N_{Re} — Reynolds number = $\rho VD/\mu$.

N_{Pr} — Prandtl number = $\mu C_p/k$.

N_{Nu} — Nusselt number = hD/k.

N^*_{Nu} — supercritical Nusselt number - hD/k.

n — number of tube banks.

n_s — number of turbine stages.

n_e — number of turbine exhaust ends.

P — pressure.

P_c — critical pressure.

P_o — dead state pressure.

$\mathcal{P}$ — power.

Q — heat flow.

Q_{rej} — rejected heat.

R, $\mathcal{R}$ — gas constant.

r — pressure ratio in expansion or compression.

S — entropy.

T — temperature.

T_c — critical temperature.

T_o — dead state heat rejection temperature.

T^* — optimum resource temperature.

U, U_o, U_{eq} — overall heat transfer coefficient.

V — velocity.

V_c— sonic velocity.

$\dot{V}$ — volumetric flow.

$\dot{V}_i$ — i^{th} stage volumetric flow.

v — specific volume.

v_c — critical specific volume.

W — component work.

W_{net} — net cycle work output.

Z_c — critical compressibility factor.

Greek Symbols

α — specific pressure ratio—$(P/P_o)^{(\gamma-1)/\gamma}$.

γ — C_p/C_v — heat capacity ratio.

δW — differential change in work.

δQ — differential change in heat flow.

ΔB — change in availability.

ΔH — change in enthalpy.

ΔH_i — i^{th} stage enthalpy or heat drop.

ΔP — change in pressure.

ΔS — change in entropy.

ΔT — change in temperature.

η_{cycle}, η_c — cycle efficiency.

η_{cf} — condenser pumping or fan related efficiency = W_C/Q_{HE}.

η_p — pump efficiency.

η_u — utilization efficiency.

η_t — turbine efficiency.

μ — viscosity.

χ — mass fraction of liquid.

π — turbine similarity parameter.

ρ — density.

ξ — figure of merit for turbine size.

Φ — component cost.

Subscripts and Superscripts

C,c — condenser or heat rejection conditions.

c — critical conditions.

cf — condenser pumping or fan related.

E — equipment.

ex — turbine exhaust conditions.

g — vapor or gas.

gf — geothermal fluid.

HE — heat exchanger.

I — indirect costs.

ℓ — liquid.

P,p — pump.

r — reduced coordinates.

sat — saturated or equilibrium conditions.

T,t — turbine.

vap — vaporization.

W,w — wells.

wf — working fluid.

10. REFERENCES

1. J.H. Anderson, "The Vapor-Turbine Cycle for Geothermal Power Production," in *Geothermal Energy*, (Stanford Univ. Press, Stanford, CA,1973).

2. D. Aronson, "Binary Cycle for Power Generation," *Proc. of Amer. Power Conf.*, 23 (1961).

3. J.C.S. Chou, "Regenerative Vapor-Turbine Cycle for Geothermal Power Plant," *Geothermal Energy* **2**, 21 (1973).

4. D.H. Cortez, B. Holt, and A.J.L. Hutchinson, "Advanced Binary Cycles for Geothermal Power Generation," *Energy Sources* **1**(1), 74 (1973).

5. A. Hansen, "Thermal Cycles for Geothermal Sites and Turbine Installation at The Geysers Power Plant, California," in Geothermal Energy, Proc., U.N. Conf. on New Sources of Energy, Rome, **3**, August 21-31, 1961, pp. 365-379.

6. V.K. Jonsson, A.J. Taylor, and A.D. Charmichael, "Optimization of Geothermal Power Plant by Use of Freon Vapour Cycle," *Timarit- VF1*, p. 2 (1969).

7. V.N. Moskvicheva, "Geopower Plant on the Paratunka River," USSR Academy of Sciences (1971).

8. P. Kruger and C. Otte, Eds., *Geothermal Energy*, (Stanford Univ. Press, Stanford, CA, 1973).

9. A.L. Austin, G.H. Higgins, and J.H. Howard, "The Total Flow Concept for Recovery of Energy from Geothermal Hot Brine Deposits," Lawrence Livermore Laboratory report UCRL-51366, Livermore, CA (April 1973).

10. D.W. Brown, M.C. Smith, and R.M. Potter, "A New Method for Extracting Energy from 'Dry' Geothermal Reservoirs," Proc. Twentieth Annual Southwestern Petroleum Short Course, Texas Tech Univ., Lubbock, Texas, April 26-27, 1973, pp. 249-255.

11. D.E. White and D.L. Williams, Eds., "Assessment of Geothermal Resources of the United States — 1975," Geological Survey Circular 726, U.S. Geological Survey, Reston, VA (1975).

12. D.E. White, "Geothermal Energy," Geological Survey Circular 519, U.S. Geological Survey, Reston, VA (1965).

13. B.F. Grossling, "An Appraisal of the Prospects of Geothermal Energy in the United States," in U.S. Energy Outlook: Natl. Petroleum Council, Washington, D.C., Chap. 2, p. 15-16 (1972).

14. R.W. Rex and D.J. Howell, "Assessment of U.S. Geothermal Resources," in *Geothermal Energy*, P. Kruger and C. Otte, Eds., (Stanford Univ. Press, Stanford, CA, 1973), pp. 59-67.

15. W.J. Hickel, "Geothermal Energy", Geothermal Resources Research Conference Final Report, Univ. of Alaska Press, Fairbanks, Alaska (1972).

16. National Petroleum Council, "U.S. Energy Outlook — New Energy Forms", U.S. Govt. Printing Office, Washington, D.C. (1973).

17. D.N. Anderson and R.G. Bowen, Eds., "Proceedings: Workshop on Environmental Aspects of Geothermal Resources Development," prepared by State of Calif. Dept. of Conservation, Div. of Oil and Gas and State of Oregon, Dept. of Geology and Mineral Industries in association with NSF/RANN, grant No. AER-75-06872, (November 1974).

18. R.G. Bowen, "Environmental Impact of Geothermal Development," in *Geothermal Energy*, P. Kruger and C. Otte, Eds., (Stanford Univ. Press, Stanford, CA, 1973), pp. 197-216.

19. R.C. Axtmann, "Environmental Impact of a Geothermal Power Plant," *Science* **187** (4179), 795 (1975).

20. "Draft Environment Impact Statement for the Geothermal Leasing Program," U.S. Department of the Interior, Washington, D.C. (September 1971).

21. S.L.Milora, "Application of the Martin Equation of State to the Thermodynamic Properties of Ammonia," Oak Ridge National Laboratory report ORNL-TM-4413, Oak Ridge, TN (December 1973).

22. J.J. Martin, "Equations of State," *Ind. Eng. Chem.* **59**, 34-52 (1967).

23. H.P. Meissner and R. Seferian, "P-V-T Relations of Gases," *Chem. Eng. Progr.* **47**, 579 (1951).

24. A.L. Lydersen, R.A. Greenkorn, and O.A. Hougen, "Generalized Thermodynamic Properties of Pure Fluids," Engineering Exp. Station, Report No. 4, Univ. of Wisconsin, Madison, WI (October 1955).

25. R.C. Reid and T.K. Sherwood, *The Properties of Gases and Liquids*, 2nd Ed., (McGraw-Hill, New York, 1966).

26. O.A. Hougen, K.M. Watson, R.A. Ragatz, *Chemical Process Principles, Part II Thermodynamics, 2nd Ed.*, (J. Wiley and Sons, New York, 1974).

27. American Society of Heating, Refrigerating, and Air-Conditioning Engineers, *ASHRAE Thermodynamic Properties of Refrigerants,* (New York, 1969).

28. "Thermodynamic Properties of Freon 22," DuPont, Organic Chemicals Department, "Freon" Products Division, Technical Publ. T-22.

29. P.R. Malbrunot, P.A. Meunier, G.M. Scatena, W.H. Mears, K.P. Murphy, and J.V. Sinka, "Pressure-Volume-Temperature Behavior of Difluoromethane," *J. Chem. Eng. Data* **13**, 16 (1968).

30. J.J. Martin, "Thermodynamic Properties of Dichlorotetrafluoroethane," *J. Chem. Eng. Data* **5**, 334 (1960).

31. T.E. Morsy, "A New Vapor Chart for Refrigerant R-114," *Kältetechnik* **3**, 86 (1965).

32. W.H. Mears, E. Rosenthal, and J.V. Sinka, "Pressure-Volume-Temperature Behavior of Pentafluoromonochloroethane," *J. Chem. Eng. Data* **11**, 338 (1966).

33. H.J. Löffler and H. Matthias, "Thermodynamic Properties of Pentafluoromonochloroethane (R-115)," *Kaltetechnik-Klimatisierung* **11**, 408 (1966).

34. L.N. Canjar and F.S. Manning, *Thermodynamic Properties and Reduced Correlations for Gases*, (Gulf Publ. Co., Houston, TX, 1967).

35. J.J. Martin, "Thermodynamic Properties of Perfluorocyclobutane," *J. Chem. Eng. Data* **7**, 68 (1962).

36. *U.K. Steam Tables in SI Units 1970*, (E. Arnold, London, 1970), p. 142.

37. J.H. Keenan and R.G. Keyes, *Steam Tables*, (J. Wiley and Sons, New York, 1968).

38. K.R. Landgraf, K.I. Kudrnac, and R. Solares, "Choice of Working Fluid and Operating Conditions for Energy Conversion with Geothermal Heat Sources," Oak Ridge National Laboratory report ORNL-MIT-180, Oak Ridge, TN (October 1973).

39. S. Troulakis, "Steam Ammonia Power Cycle Study," Report No. STA-19, DeLaval Turbine Inc., Trenton, N.J. (1968).

40. O.E. Baljé, "A Study on Design Criteria and Matching of Turbomachines, Parts A & B," *J. Eng. for Power*, Trans. ASME: 83 (January 1962).

41. R.G. Seth and W. Steiglemann, "Binary-Cycle Power Plants Using Dry Cooling Systems, Part I, Technical and Economic Evaluation," The Franklin Institute Research Laboratories final report F-C3023, Phila., PA (January 1972).

42. L. Heller, "New Power Station System for Unit Capacities in the 1000 MW Order," *Acta. Techn. Hung.* **50**, 93-123 (1965).

43. A. Lavi, Ed., "Proceedings of Solar Sea Power Plant Conference and Workshop," Carnegie Mellon Univ., Pittsburgh, PA (June 1973).

44. A. Lavi, "Solar Sea Project," Report NSF/RANN/SE/GI-39114/PR/74/6, Carnegie Mellon Univeristy, Pittsburgh, PA (January 1975).

45. M.S. Peters and K.D. Timmerhaus, *Plant Design and Economics for Chemical Engineers*, 2nd Ed., (McGraw-Hill, New York, 1968).

46. D.F. Rudd and C.C. Watson, *Strategy of Process Engineering*, (J. Wiley & Sons, New York, 1968).

47. J.H. Altseimer, "Geothermal Well Technology and Potential Applications of Subterrene Devices — A Status Review," Los Alamos Scientific Laboratory report LA-5689-MS, Los Alamos, NM (August 1974).

48. R. Greider, "Economic Considerations for Geothermal Exploration in the Western United States," presented at the Symposium of Colorado Department of Natural Resources, Denver, Colorado (December 1973).

49. R.H. Perry and C.H. Chilton, *Chemical Engineers Handbook, 5th Ed.*, (McGraw-Hill, New York, 1973).

50. W.H. McAdams, *Heat Transmission, 3rd Ed.*, (McGraw-Hill, New York, 1974).

51. R.P. Bringer and J.M. Smith, "Heat Transfer in the Critical Region," *AIChE J.* **3** (1), 49 (1957).

52. V.S. Sastry, "An Analytical Investigation of Forced Convection Heat Transfer to Fluids Near the Thermodynamic Critical Point," Amer. Soc. Mech. Eng. paper ASME 74-WA/HT-29 (November 1974).

53. A.P. Fraas and M.N. Ozisik, *Heat Exchanger Design*, (J. Wiley and Sons, New York, 1965).

54. P.L. Jensen, P.C. Ahrens, and A.S.Y. Ho, "Utilization of a Low Temperature (300°F) Water-Dominated Geothermal Source for Power Generation," Oak Ridge National Laboratory report ORNL-MIT-187, Oak Ridge, TN (March 1974).

55. K.E. Nichols, "Turbine Prime Mover Cost Model Empirical Equation, P.O. 1/Y-49428V," Barber-Nichols Engineering Co., Arvada, CO (March 1975).

56. C.H. Bloomster, P.D. Cohn, J.G. DeSteese, H.D. Huber, P.N. LaMori, D.W. Shannon, J.R. Sheff, and R.A. Walter, "GEOCOST: A Computer Program for Geothermal Cost Analysis," Battelle Northwest Laboratory report BNWL-1888, UC-13 Richland, WA (February 1975).

57. R.A. Walter, C.M. Bloomster and S.E. Wise, "Thermodynamic Modelling of Geothermal Power Plant," Battelle Northwest Laboratory, report BNWL-1911, UC-13 Richland, WA (November 1975).

58. M.A. Green and H.S. Pines, "Calculation of Geothermal Power Plant Cycles Using Program GEOTHM," paper No. VII-12, Second United Nations Geothermal Energy Symposium, San Francisco, CA, May, 1975.

59. R.M. Potter, "Assessment of Some of the Geothermal Resources of the Eastern United States," paper presented at the Near-Normal Geothermal Gradient Workshop, U.S. Energy Research and Development Administration, Washington, D.C., March 10-11, 1975.

60. D.W. Brown, "The Potential for Hot-Dry-Rock Geothermal Energy in the Western United States," Hearings before the Subcommittee on Energy of the Committee on Science and Astronautics, U.S. House of Representatives, Ninety-Third Congress, First Session on H.R. 8628, H.R. 9658, Sept. 11, 13, and 18, 1973, pp. 129-138.

61. T. Meidav, "Geothermal Opportunities Bear a Closer Look," *Oil and Gas J.* **72**, 102 (1974).

62. T. Meidav, "Time is of the Essence in Developing Geothermal Energy,"*Oil and Gas J.* **73,** 168 (1975).

63. I.C. Bupp, J.C. Devian, M-P. Donsimoni, and R. Treitel, "The Economics of Nuclear Power," *Technology Review* **77** (4), 14 (1975).

64. A. Sesonske, "*Nuclear Power Plant Design Analysis,* (U.S. Atomic Energy Commission report TID-26241, Nat. Tech. Info. Service, Springfield, VA, 1973).

65. H.C. Hottel and J.B. Howard, *New Energy Technology* (MIT Press, Cambridge, MA,1971).

66. R. Cataldi, P. DiMario, and T. Leardini, "Application of Geothermal Energy to the Supply of Electricity in Rural Areas," *Geothermics* **2** (1), 3 (1973).

67. S.E. Beall and G. Samuels, "The Use of Warm Water for Heating and Cooling Plant and Animal Enclosures," Oak Ridge National Laboratory report ORNL-TM-3381, Oak Ridge, TN (1971).

68. S.E. Beall, "Agricultural and Urban Uses of Low-Temperature Heat," in *Beneficial Uses of Thermal Discharges*, S.P. Mathur and R. Stewart, Eds. (New York State Department of Environmental Conservation, Albany, NY, 1970), pp. 185-202.

69. S.E. Beall, "Waste Heat Uses Cut Thermal Pollution," *Mech. Eng.* **93**, 15 (1971).

70. G. Bodvarsson, L. Boersma, R. Couch, L. David, and G. Reistad, "Systems Study for the Use of Geothermal Energies in the Pacific Northwest," Oregon State University report RLO-2227-T19-1, Corvallis, OR (1974), and G.M. Reistad, "Analysis of Potential Nonelectrical Applications of Geothermal Energy and Their Place in the National Economy," Lawrence Livermore Laboratory report UCRL-51747, Livermore, CA (February 1975).

71. 1972 Joint Association Survey of the U.S. Oil and Gas Producing Industry, "Section I. Drilling Costs, Section II, Expenditures for Exploration, Development and Production," (November 1973).

72. 1973 Joint Association Survey of the U.S. Oil and Gas Producing Industry, "Section I, Drilling Costs," (February 1975).

73. E.M. Bee Dagum and K-P. Heiss, "An Econometric Study of Small and Intermediate Size Diameter Drilling Costs for the United States," PNE-3012 (Vol. 1 and 2) prepared for the U.S. Atomic Energy Commission by Mathematica, Princeton, N.J. (June 1968).

74. E.M. Shoemaker, Ed., "Continental Drilling," report of the Workshop on Continental Drilling, Abiquiu, New Mexico, by the Carnegie Institution, Washington, D.C. (June 1975).

75. R. Hendron, personal communication, Los Alamos Scientific Laboratory, Los Alamos, N.M. September 1975.

76. J.L. Hair, "Construction Techniques and Costs for Underground Emplacement of Nuclear Explosives," U.S. Army Corps of Engineers report PNE-5004F, Fort Worth, TX (April 1969).

77. M. Nathenson and L.J.P. Muffler, "Geothermal Resources in Hydrothermal Convection Systems and Conduction-Dominated Areas," in *Assessment of Geothermal Resources of the United States — 1975,"* D.E. White and D.L. Williams, Eds. Geological Survey Circular 726, U.S. Geological Survey, Reston, VA (1975).

78. J.R. McNitt, "Exploration and Development of Geothermal Power in California," special report 75, Calif. Div. of Mines and Geology, San Francisco, CA (1967).

79. D. Towse, "Estimate of Geothermal Energy Resources in the Salton Trough, California," Lawrence Livermore Laboratory report UCRL-51851, Berkeley, CA (June 1975).

80. R.D. McFarland, "Geothermal Reservoir Models — Crack Plane Model," Los Alamos Scientific Laboratory report LA-5947-MS, Los Alamos, NM (April 1975).

81. F.H. Harlow and W.E. Pracht, "A Theoretical Study of Geothermal Energy Extraction,"*J. of Geophysical Research* **77**, 7038 (1972).

82. T.A. Fearnside and F.C. Cheney, "Fast Estimate of Power Plant Costs," ASME paper, Power and Management Divisions, New York (1972).

83. S.L. Milora, "STATEQ: A Nonlinear Least-Squares Code for Obtaining Martin Thermodynamic Representations of Fluids in the Gaseous and Dense Gaseous Regions," Oak Ridge National Laboratory report ORNL-TM-5115, Oak Ridge, TN (to be published).

INDEX

Dat

Retu